Terrence James Victorino (Ed.)

Nokia Asha 302

Terrence James Victorino (Ed.)

Nokia Asha 302

Series 40, Nokia, Wi-Fi

Log Press

Publisher:
Log Press is a trademark of
International Book Market Service Ltd., 17 Rue Meldrum, Beau Bassin, 1713-01 Mauritius
Email: info@bookmarketservice.com
Website: www.bookmarketservice.com

Published in 2012

Printed in: U.S.A., U.K., Germany. This book was not produced in Mauritius.

ISBN: 978-620-1-70806-8

Contents

Articles

References

Nokia_Asha_302

Nokia Asha 302

Manufacturer	Nokia
Series	Touch & Type
Compatible networks	• GSM 850 / 900 / 1800 / 1900 • GPRS/EDGE class B, multislot class 33 • UMTS 850 / 900 / 1700 / 1900 / 2100 • HSDPA Cat9, 10.2 Mbps • HSUPA Cat5 2.0 Mbps
Introductory price	€95 [1]
Availability by country	Global
Form factor	Qwerty candybar
Dimensions	• Width: 55.7 mm • Height: 116.5 mm • Thickness: 13.9 mm
Weight	99 g
Operating system	Nokia Asha Series 40 OS
CPU	• 1 GHz ARM11
Memory	• 128 MB RAM • 256 MB ROM
Storage	• 170 MB internal NAND memory (100MB available to end user)
Removable storage	up to 32 GB microSDHC
Battery	• BP-3L 1300 mAh Li-Ion battery (removable) • micro USB and 2 mm DC plug charging
Data inputs	• QWERTY keyboard • External functional hardware keys
Display	transmissive LCD 320 x 240 px (QVGA), 2.4" (66 mm), 18 bits
Rear camera	3.2 MP (CMOS sensor) EDoF
Front camera	No
Ringtones & notifications	Nokia Tune
Connectivity	• WLAN IEEE 802.11 b/g/n (2.4 GHz) • bluetooth 2.1 +EDR • micro USB 2.0 • USB On-the-Go 1.3 • 3.5 mm AV connector (audio in/out) • SIM card • FM receiver with RDS

The **Nokia Asha 302** is phone powered by Nokia's Series 40 operating system. It was announced at Mobile World Congress 2012 in Barcelona along with other Asha phones - the Nokia Asha 202, and 203. The 302 is considered to be among the flagship of the Asha family. Its main features are the QWERTY keyboard, the pentaband 3G radio, SIP VoIP over 3G and Wi-Fi. The software update of Nokia Asha 302 is Mail for Exchange supported.[2]

History and availability

The Nokia Asha 302 was announced at Mobile World Congress 2012 in Barcelona. It will be available shortly Globally. The phone will be sold at a price of €95 subject to taxes and subsidies.

Hardware

Processors

The Nokia Asha 302 is powered by the same 1 GHz ARM11 processor found in some Symbian^3 phones such as the Nokia 500, 600 and 700 but lack the dedicated Broadcom GPU which is not supported by the Nokia Series 40 operating system. The system also has 128 MB of low power single channel RAM (Mobile DDR).

Screen and input

The Nokia Asha 302 has a 2.4-inch transmissive LCD screen with a resolution of 320 × 240 pixel. In contrast with the Nokia Asha 303, the screen of the Asha 302 is wider than taller. According to Nokia it is capable of displaying up to 262 thousands colors. The device also has a backlit 4-row keyboard with regional variant available (QWERTY, AZERTY, etc. ...).

The back camera has an Extended Depth of Field (EDoF) feature (no mechanical zoom), no flash and has a 4× digital zoom for both video and camera. The sensor size of the back camera is 3.2 megapixel (2048 x 1536 px), has a f/2.8 aperture and a 50 cm to infinity focus range. It is capable of video recording at up to 640 x 480 px at 15 fps with mono sound.

Audio and output

The Nokia Asha 302 has one microphone and a loudspeaker, which is situated on the back of the device. On the top, there is a 3.5 mm AV connector which simultaneously provides stereo audio output and microphone input. Between the 3.5 mm AV connector and the 2 mm charging connector, there is a High-Speed USB 2.0 USB Micro AB connector provided for data synchronization, battery charging and supports for USB On-The-Go 1.3 (the ability to act as a USB host) using a Nokia Adapter Cable for USB OTG CA-157 (not included upon purchase).

The built-in Bluetooth v2.1 +EDR (Enhanced Data Rate) supports stereo audio output with the A2DP profile. Built-in car hands-free kits are also supported with the HFP profile. File transfer is supported (FTP) along with the OPP profile for sending/receiving objects. It is possible to remote control the device with the AVRCP profile. It supports wireless earpieces and headphones through the HSP profile. The DUN profile which permits access to the Internet from a laptop by dialing up on a mobile phone wirelessly (tethering) and PAN profile for networking using Bluetooth are also supported. The device also functions as a FM Receiver, allowing one to listen to the FM radio by using headphones connected to the 3.5 jack as antenna.

Battery and SIM

The battery life of the BL-5J (1430 mAh) as claimed by Nokia is from 6 to 9 hours of talk time, from 29 to 34 days of standby and 50 hours of music playback depending on actual usage.

The SIM card is located under the battery which can be accessed by removing the back panel of the device. The microSDHC card socket is also located under the back cover (but not under the battery). No tool is necessary to remove the back panel.

Storage

The phone has 150 MB of available non-removable storage. Additional storage is available via a hot swappable microSDHC card socket, which is certified to support up to 32 GB of additional storage.

Software

The Nokia Asha 302 is powered by Nokia Series 40 operating system with service pack 1 and comes with a variety of applications:

- **Web**: Nokia (proxy) Browser for Series 40
- **Conversations**: Nokia Messaging Service 3.2 (instant messaging and e-mail) and SMS, MMS
- **Social**: Facebook, Twitter, Flickr and Orkut
- **Media**: Camera, Photos, Music player, Nokia Music Store (on selected market), Flash Lite 3.0 (for YouTube video), Video player
- **Personal Information Management**: Calendar, Detailed contact information
- **Utilities**: VoIP, Notes, Calculator, To-do list, Alarm clock, Voice recorder, Stopwatch
- **Games**: *Angry Birds Lite* (first level only, additional levels can be purchased on the Nokia Store)

The Home screen is customizable and allow the user to add, amongst others, favorite contacts, Twitter/Facebook feeds, applications shortcuts, IM/e-mail notifications and calendar alert.

References

[1] http://press.nokia.com/2012/02/27/nokia-expands-its-asha-range-with-smarter-feature-phones-that-improve-ways-to-work-learn-and-play/

[2] "VoIP support in Nokia devices - Nokia Developer Wiki" (http://www.developer.nokia.com/Community/Wiki/VoIP_support_in_Nokia_devices#Support_in_Series_40_devices). Nokia. . Retrieved 15 December 2011.

External links

- http://www.nokia.com/nokia-asha-smarter-mobile-phones
- http://europe.nokia.com/find-products/devices/nokia-asha-302/specifications
- http://www.developer.nokia.com/Devices/Device_specifications/302
- http://www.developer.nokia.com/Community/Wiki/VoIP_support_in_Nokia_devices#Support_in_Series_40_devices
- http://a.com.pk/djuice-nokia-302-smartphone-with-free-mobile-internet/
- Nokia Asha 302 Review (http://www.reviewsexpert.net/2012/02/nokia-asha-302-review.html)

Series_40

Series 40 is a software platform and application user interface (UI) software on Nokia's broad range of mid-tier feature phones, as well as on some of the Vertu line of luxury phones. It is the world's most widely used mobile phone platform and found in hundreds of millions of devices.[1] Nokia announced on 25 January 2012 that the company has sold over 1.5 billion S40 devices. [2] S40 has more features than the Series 30, which is a very basic OS. Neither of them are based on Symbian.

Series 40-based Nokia 6300

History

Series 40 was officially introduced in 1999 with the release of the Nokia 7110. It had a 96 × 65 pixel monochrome display and was the first phone to come with a WAP browser. Over the years, the S40 UI has evolved from a low resolution UI to a high resolution color UI with an enhanced graphical look. The third generation of Series 40 that became available in 2005 introduced support for devices with resolutions as high as QVGA (240×320).[3] It is possible to customize the look-and-feel of the UI via comprehensive themes.[4] A list of all Series 40 devices can be found on the Nokia web site.[5]

In 2012, the new Nokia Asha mobile phones 200/201, 302, 303 and 311 were released and all use Series 40.[5]

Home screen of Nokia S40 v10.80 running on Nokia 6303i

Features

Applications

It provides communication applications such as telephone, internet telephony (VoIP), messaging, email client with POP3 and IMAP4 capabilities and Web browser; media applications such as camera, video recorder, music/video player and FM radio; and phonebook and other personal information management (PIM) applications such as calendar and tasks. Basic file management, like in Series 60, is provided in the Applications and Gallery folders and subfolders. Gallery is also the default location for files transferred over Bluetooth to be placed. User-installed applications on Series 40 are generally mobile Java applications. Flash Lite applications are also supported, but mostly used for screensavers.[6]

Web browser

The integrated web browser can access most web content through the service provider's XHTML/HTML gateway. The latest version of Series 40, called Series 40 6th Edition, introduced a new browser based on the WebKit open source components WebCore and JavaScriptCore. The new browser delivers support for HTML 4.01, CSS2, JavaScript 1.5, and Ajax. Also, like the higher-end Series 60, Series 40 can run the Opera Mini web browser to enhance the user's web browsing experience.

Synchronization

Support for SyncML synchronization with external services of the address book, calendar and notes is present. However with many S40 phones, these synchronization settings must be sent via an OTA text message.

Technical

Software platform

Series 40 is an embedded software platform that is open for software development via standard or de-facto content and application development technologies. It supports Java MIDlets, i.e. Java MIDP and CLDC technology, which provide location, communication, messaging, media, and graphics capabilities.[7] S40 also supports Flash Lite applications.[6]

Operating system

Series 40 is a simpler operating system than the higher end S60. Because S40 devices do not support true multi-tasking and do not have a native code API for third parties, its user interface may appear to be more responsive and faster than the other Nokia platforms Symbian S60 on similar hardware.[8]

References

[1] "Forum Nokia - Nokia Series 40 Platform" (http://www.forum.nokia.com/Devices/Series_40/). Nokia. . Retrieved 2010-10-27.
[2] "Nokia has sold over 1.5 billion Series 40 phones" (http://www.esphoneblog.com/2012/01/25/nokia-has-sold-over-1-5-billion-series-40-phones/). . Retrieved 2012-01-25.
[3] "Series 40 UI Style Guide – Forum Nokia" (http://www.forum.nokia.com/info/sw.nokia.com/id/73e935fe-8b59-43b2-ab3e-1c5f763672db/Series_40_UI_Style_Guide.html). Nokia. . Retrieved 2008-09-26.
[4] "Carbide.ui Theme Edition (can be used to create S40 themes) – Forum Nokia" (http://www.forum.nokia.com/Library/Tools_and_downloads/Other/Carbide.ui/). Nokia. . Retrieved 2011-05-16.
[5] "Device specifications, filtered for Series 40" (http://www.developer.nokia.com/Devices/Device_specifications/?filter1=s40). Nokia. . Retrieved 2012-06-01.
[6] "Working with Nokia Series 40 Flash Lite content – Adobe Developer Center" (http://web.archive.org/web/20080518220200/http://www.adobe.com/devnet/devices/articles/nokia_series40_pt1_print.html). Adobe Systems. Archived from the original (http://www.adobe.com/devnet/devices/articles/nokia_series40_pt1_print.html) on 2008-05-18. . Retrieved 2008-09-26.
[7] "Developing Scalable Series 40 Applications, A Guide for Java Developers" (http://www.pearsoned.co.uk/BOOKSHOP/detail.asp?item=100000000075919). Addison-Wesley. . Retrieved 2008-09-26.
[8] "Comparing Series 40 against S60 (as of 2007) – All about Symbian" (http://www.allaboutsymbian.com/features/item/Series_40_vs_S60.php). . Retrieved 2008-09-26.

Nokia

Nokia Corporation

NOKIA	
Type	Public
Traded as	OMX: NOK1V [1], NYSE: NOK [2], FWB: NOA3 [3]
Industry	Telecommunications equipment Internet Computer software
Founded	Tampere, Grand Duchy of Finland (1865) incorporated in Nokia (1871)
Founder(s)	Fredrik Idestam Leo Mechelin
Headquarters	Espoo, Finland
Area served	Worldwide
Key people	Risto Siilasmaa (Chairman) Stephen Elop (President & CEO)
Products	Mobile phones Smartphones Mobile computers Networks (See products listing)
Services	Maps and navigation, music, messaging and media Software solutions (See services listing)
Revenue	▼ €38.65 billion (2011)[4]
Operating income	▼ € -1.073 billion (2011)[4]
Net income	▼ € -1.164 billion (2011)[4]
Total assets	▼ €36.20 billion (2011)[4]
Total equity	▼ €11.87 billion (2011)[4]
Employees	122,148 (2012)[5]
Divisions	Mobile Solutions Mobile Phones Markets
Subsidiaries	Nokia Siemens Networks Navteq Vertu Qt Development Frameworks
Website	Nokia.com [6]

Nokia Corporation (Finnish pronunciation: [ˈnokiɑ], English /ˈnɒkiə/) (OMX: NOK1V [1], NYSE: NOK [2], FWB: NOA3 [3]) is a Finnish multinational communications and information technology corporation headquartered in Keilaniemi, Espoo, Finland.[7] Its principal products are mobile telephones and portable IT devices. It also offers Internet services including applications, games, music, maps, media and messaging through its Ovi platform, and free-of-charge digital map information and navigation services through its wholly owned subsidiary Navteq.[8] Nokia has a joint venture with Siemens, Nokia Siemens Networks, which provides telecommunications network equipment and services.[9]

Nokia has around 122,000 employees across 120 countries, sales in more than 150 countries and annual revenues of around €38 billion.[4] As of 2012 it is the world's second-largest mobile phone maker by unit sales (after Samsung), with a global market share of 22.5% in the first quarter.[10] Nokia is a public limited-liability company listed on the Helsinki, Frankfurt, and New York stock exchanges.[11] It is the world's 143rd-largest company measured by 2011 revenues according to the *Fortune Global 500*.[12]

Nokia was the world's largest vendor of mobile phones from 1998 to 2012.[10] However, over the past five years it has suffered declining market share as a result of the growing use of smartphones, principally the Apple iPhone and devices running on Google's Android operating system. As a result, its share price has fallen from a high of US$40 in 2007 to under US$3 in 2012.[13][14] Since February 2011, Nokia has had a strategic partnership with Microsoft, as part of which all Nokia smartphones will incorporate Microsoft's Windows Phone operating system (replacing Symbian). Nokia unveiled its first Windows Phone handsets, the Lumia 710 and 800, in October 2011.[15]

History

Pre-telecommunications era

Fredrik Idestam, co-founder of Nokia.

Statesman Leo Mechelin, co-founder of Nokia.

The predecessors of the modern Nokia were the Nokia Company (Nokia Aktiebolag), Finnish Rubber Works Ltd (Suomen Gummitehdas Oy) and Finnish Cable Works Ltd (Suomen Kaapelitehdas Oy).[16]

Nokia's history started in 1865 when mining engineer Fredrik Idestam established a groundwood pulp mill on the banks of the Tammerkoski rapids in the town of Tampere, in southwestern Finland in the Russian Empire and started manufacturing paper.[17] In 1868, Idestam built a second mill near the town of Nokia, fifteen kilometres (nine miles) west of Tampere by the Nokianvirta river, which had better resources for hydropower production.[18] In 1871, Idestam, with the help of his close friend statesman Leo Mechelin, renamed and transformed his firm into a share company, thereby founding the Nokia Company, the name it is still known by today.[18]

Toward the end of the 19th century, Mechelin's wishes to expand into the electricity business were at first thwarted by Idestam's opposition. However, Idestam's retirement from the management of the company in 1896 allowed Mechelin to become the company's chairman (from 1898 until 1914) and sell most shareholders on his plans, thus realizing his vision.[18] In 1902, Nokia added electricity generation to its business activities.[17]

Industrial conglomerate

In 1898, Eduard Polón founded Finnish Rubber Works, manufacturer of galoshes and other rubber products, which later became Nokia's rubber business.[16] At the beginning of the 20th century, Finnish Rubber Works established its factories near the town of Nokia and they began using Nokia as its product brand.[19] In 1912, Arvid Wickström founded Finnish Cable Works, producer of telephone, telegraph and electrical cables and the foundation of Nokia's cable and electronics businesses.[16] At the end of the 1910s, shortly after World War I, the Nokia Company was nearing bankruptcy.[20] To ensure the continuation of electricity supply from Nokia's generators, Finnish Rubber Works acquired the business of the insolvent company.[20] In 1922, Finnish Rubber Works acquired Finnish Cable Works.[21] In 1937, Verner Weckman, a sport wrestler and Finland's first Olympic Gold medalist, became president of Finnish Cable Works, after 16 years as its technical director.[22] After World War II, Finnish Cable Works supplied cables to the Soviet Union as part of Finland's war reparations. This gave the company a good foothold for later trade.[22]

The three companies, which had been jointly owned since 1922, were merged to form a new industrial conglomerate, Nokia Corporation in 1967 and paved the way for Nokia's future as a global corporation.[23] The new company was involved in many industries, producing at one time or another paper products, car and bicycle tires, footwear (including rubber boots), communications cables, televisions and other consumer electronics, personal computers, electricity generation machinery, robotics, capacitors, military communications and equipment (such as the SANLA M/90 device and the M61 gas mask for the Finnish Army), plastics, aluminium and chemicals.[24] Each business unit had its own director who reported to the first Nokia Corporation President, Björn Westerlund. As the president of the Finnish Cable Works, he had been responsible for setting up the company's first electronics department in 1960, sowing the seeds of Nokia's future in telecommunications.[25]

Eventually, the company decided to leave consumer electronics behind in the 1990s and focused solely on the fastest growing segments in telecommunications.[26] Nokian Tyres, manufacturer of tires, split from Nokia Corporation to form its own company in 1988[27] and two years later Nokian Footwear, manufacturer of rubber boots, was founded.[19] During the rest of the 1990s, Nokia divested itself of all of its non-telecommunications businesses.[26]

Telecommunications era

The seeds of the current incarnation of Nokia were planted with the founding of the electronics section of the cable division in 1960 and the production of its first electronic device in 1962: a pulse analyzer designed for use in nuclear power plants.[25] In the 1967 fusion, that section was separated into its own division, and began manufacturing telecommunications equipment. A key CEO and subsequent Chairman of the Board was *vuorineuvos* Björn "Nalle" Westerlund (1912–2009), who founded the electronics department and let it run at a loss for 15 years.

Networking equipment

In the 1970s, Nokia became more involved in the telecommunications industry by developing the Nokia DX 200, a digital switch for telephone exchanges. The DX 200 became the workhorse of the network equipment division. Its modular and flexible architecture enabled it to be developed into various switching products.[28] In 1984, development of a version of the exchange for the Nordic Mobile Telephony network was started.[29]

For a while in the 1970s, Nokia's network equipment production was separated into *Telefenno*, a company jointly owned by the parent corporation and by a company owned by the Finnish state. In 1987, the state sold its shares to Nokia and in 1992 the name was changed to Nokia Telecommunications.

In the 1970s and 1980s, Nokia developed the Sanomalaitejärjestelmä ("Message device system"), a digital, portable and encrypted text-based communications device for the Finnish Defence Forces.[30] The current main unit used by the Defence Forces is the Sanomalaite M/90 (SANLA M/90).[31]

First mobile phones

The Mobira Cityman 150, Nokia's NMT-900 mobile phone from 1989 (left), compared to the Nokia 1100 from 2003.[32] The Mobira Cityman line was launched in 1987.[33]

The technologies that preceded modern cellular mobile telephony systems were the various "0G" pre-cellular mobile radio telephony standards. Nokia had been producing commercial and some military mobile radio communications technology since the 1960s, although this part of the company was sold some time before the later company rationalization. Since 1964, Nokia had developed VHF radio simultaneously with Salora Oy. In 1966, Nokia and Salora started developing the ARP standard (which stands for Autoradiopuhelin, or *car radio phone* in English), a car-based mobile radio telephony system and the first commercially operated public mobile phone network in Finland. It went online in 1971 and offered 100% coverage in 1978.[34]

In 1979, the merger of Nokia and Salora resulted in the establishment of Mobira Oy. Mobira began developing mobile phones for the NMT (Nordic Mobile Telephony) network standard, the first-generation, first fully automatic cellular phone system that went online in 1981.[35] In 1982, Mobira introduced its first car phone, the Mobira Senator for NMT-450 networks.[35]

Nokia bought Salora Oy in 1984 and now owning 100% of the company, changed the company's telecommunications branch name to Nokia-Mobira Oy. The Mobira Talkman, launched in 1984, was one of the world's first transportable phones. In 1987, Nokia introduced one of the world's first handheld phones, the Mobira Cityman 900 for NMT-900 networks (which, compared to NMT-450, offered a better signal, yet a shorter roam). While the Mobira Senator of 1982 had weighed 9.8 kg (**unknown operator: u'strong'** lb) and the Talkman just under 5 kg (**unknown operator: u'strong'** lb), the Mobira Cityman weighed only 800 g (**unknown operator: u'strong'** oz) with the battery and had a price tag of 24,000 Finnish marks (approximately €4,560).[33] Despite the high price, the first phones were almost snatched from the sales assistants' hands. Initially, the mobile phone was a "yuppie" product and a status symbol.[24]

Nokia's mobile phones got a big publicity boost in 1987, when Soviet leader Mikhail Gorbachev was pictured using a Mobira Cityman to make a call from Helsinki to his communications minister in Moscow. This led to the phone's nickname of the "Gorba".[33]

In 1988, Jorma Nieminen, resigning from the post of CEO of the mobile phone unit, along with two other employees from the unit, started a notable mobile phone company of their own, Benefon Oy (since renamed to GeoSentric).[36] One year later, Nokia-Mobira Oy became Nokia Mobile Phones.

Involvement in GSM

Nokia was one of the key developers of GSM (Global System for Mobile Communications),[37] the second-generation mobile technology which could carry data as well as voice traffic. NMT (Nordic Mobile Telephony), the world's first mobile telephony standard that enabled international roaming, provided valuable experience for Nokia for its close participation in developing GSM, which was adopted in 1987 as the new European standard for digital mobile technology.[38][39]

Nokia delivered its first GSM network to the Finnish operator Radiolinja in 1989.[40] The world's first commercial GSM call was made on 1 July 1991 in Helsinki, Finland over a Nokia-supplied network, by then Prime Minister of Finland Harri Holkeri, using a prototype Nokia GSM phone.[40] In 1992, the first GSM phone, the Nokia 1011, was launched.[40][41] The model number refers to its launch date, 10 November.[41] The Nokia 1011 did not yet employ Nokia's characteristic ringtone, the Nokia tune. It was introduced as a ringtone in 1994 with the Nokia 2100 series.[42]

GSM's high-quality voice calls, easy international roaming and support for new services like text messaging (SMS) laid the foundations for a worldwide boom in mobile phone use.[40] GSM came to dominate the world of mobile telephony in the 1990s, in mid-2008 accounting for about three billion mobile telephone subscribers in the world, with more than 700 mobile operators across 218 countries and territories. New connections are added at the rate of 15 per second, or 1.3 million per day.[43]

Personal computers and IT equipment

In the 1980s, Nokia's computer division Nokia Data produced a series of personal computers called MikroMikko.[44] MikroMikko was Nokia Data's attempt to enter the business computer market. The first model in the line, MikroMikko 1, was released on 29 September 1981,[45] around the same time as the first IBM PC. However, the personal computer division was sold to the British ICL (International Computers Limited) in 1991, which later became part of Fujitsu.[46] MikroMikko remained a trademark of ICL and later Fujitsu. Internationally the MikroMikko line was marketed by Fujitsu as the ErgoPro.

The Nokia Booklet 3G mini laptop.

Fujitsu later transferred its personal computer operations to Fujitsu Siemens Computers, which shut down its only factory in Espoo, Finland (in the Kilo district, where computers had been produced since the 1960s) at the end of March 2000,[47][48] thus ending large-scale PC manufacturing in the country. Nokia was also known for producing very high quality CRT and early TFT LCD displays for PC and larger systems application. The Nokia Display Products' branded business was sold to ViewSonic in 2000.[49] In addition to personal computers and displays, Nokia used to manufacture DSL modems and digital set-top boxes.

Nokia re-entered the PC market in August 2009 with the introduction of the Nokia Booklet 3G mini laptop.[50]

Challenges of growth

In the 1980s, during the era of its CEO Kari Kairamo, Nokia expanded into new fields, mostly by acquisitions. In the late 1980s and early 1990s, the corporation ran into serious financial problems, a major reason being its heavy losses by the television manufacturing division and businesses that were just too diverse.[51] These problems, and a suspected total burnout, probably contributed to Kairamo taking his own life in 1988. After Kairamo's death, Simo Vuorilehto became Nokia's Chairman and CEO. In 1990–1993, Finland underwent severe economic depression,[52] which also struck Nokia. Under Vuorilehto's management, Nokia was severely overhauled. The company responded by streamlining its telecommunications divisions, and by divesting itself of the television and PC divisions.[53]

The Nokia House, Nokia's head office located by the Gulf of Finland in Keilaniemi, Espoo, was constructed between 1995 and 1997. It is the workplace of more than 1,000 Nokia employees.[24]

Probably the most important strategic change in Nokia's history was made in 1992, however, when the new CEO Jorma Ollila made a crucial strategic decision to concentrate solely on telecommunications.[26] Thus, during the rest of the 1990s, the rubber, cable and consumer electronics divisions were gradually sold as Nokia continued to divest itself of all of its non-telecommunications businesses.[26]

As late as 1991, more than a quarter of Nokia's turnover still came from sales in Finland. However, after the strategic change of 1992, Nokia saw a huge increase in sales to North America, South America and Asia.[54] The exploding worldwide popularity of mobile telephones, beyond even Nokia's most optimistic predictions, caused a logistics crisis in the mid-1990s.[55] This prompted Nokia to overhaul its entire logistics operation.[56] By 1998, Nokia's focus on telecommunications and its early investment in GSM technologies had made the company the world's largest mobile phone manufacturer,[54] a position it would hold for the next 14 consecutive years until 2012. Between 1996 and 2001, Nokia's turnover increased almost fivefold from 6.5 billion euros to 31 billion euros.[54] Logistics continues to be one of Nokia's major advantages over its rivals, along with greater economies of scale.[57][58]

Recent history

Product releases

Nokia launched its Nokia 1100 handset in 2003,[32] with over 200 million units shipped, was the best-selling mobile phone of all time and the world's top-selling consumer electronics product.[59] Also that year, Nokia is cited in Michael Saylor's 2012 book, *The Mobile Wave: How Mobile Intelligence Will Change Everything*, as one of the first players in the mobile space to recognize that there was a market opportunity in combining a game console and a mobile phone (both of which many gamers were carrying in 2003) into the N-Gage. The N-Gage was a mobile phone and game console meant to lure gamers away from the Game Boy Advance, though it cost twice as much and was said to resemble a taco.[60]

Reduction in size of Nokia mobile phones

In May 2007, Nokia released its first touch screen phone, the Nokia 7710, which was also a huge success. In November 2007, Nokia announced and released the Nokia N82, its first Nseries phone with Xenon flash. At the Nokia World conference in December 2007, Nokia announced their "Comes With Music" program: Nokia device buyers are to receive a year of complimentary access to music downloads.[61] The service became commercially available in the second half of 2008.

Evolution of the Nokia Communicator. Models 9000, 9110, 9210, 9300 and 9500 shown.

Nokia Productions was the first ever mobile filmmaking project directed by Spike Lee. Work began in April 2008, and the film premiered in October 2008.[62]

In 2008, Nokia released the Nokia E71 which was marketed to directly compete with the other BlackBerry-type devices offering a full "qwerty" keyboard and cheaper prices. Nokia announced in August 2009 that they will be selling a high-end Windows-based mini laptop called the Nokia Booklet 3G.[50] On 2 September 2009, Nokia launched two new music and social networking phones, the X6 and X3.[63] The Nokia X6 features 32GB of on-board memory with a 3.2" finger touch interface and comes with a music playback time of 35 hours. The Nokia X3 is a first series 40 Ovi Store-enabled device. The X3 is a music device that comes with stereo speakers, built-in FM radio, and a 3.2 megapixel camera. On 10 September 2009, Nokia unveiled the 7705 Twist, a phone sporting a square shape that swivels open to reveal a full QWERTY keypad, featuring a 3 megapixel camera, web browsing, voice commands and weighting around 3.44 ounces (**unknown operator: u'strong'** g).[64] On 9 August 2012, Nokia launched for the Indian market two new Asha range of handsets equipped with cloud accelerated Nokia browser, helping users browse the Internet faster and lower their spend on data charges.[65]

Symbian

Originally Nokia phones had a custom Nokia OS operating system developed specifically for Nokia mobile phones.

The first Nseries device, the N90, utilised the older Symbian OS 8.1 mobile operating system, as did the N70. Subsequently Nokia switched to using SymbianOS 9 for all later Nseries devices (except the N72, which was based on the N70). Newer Nseries devices incorporate newer revisions of SymbianOS 9 that include Feature Packs. The N800, N810, N900, N9 and N950 are as of April 2012 the only Nseries devices (therefore excluding Lumia devices) to not use Symbian OS. They use the Linux-based Maemo.[66]

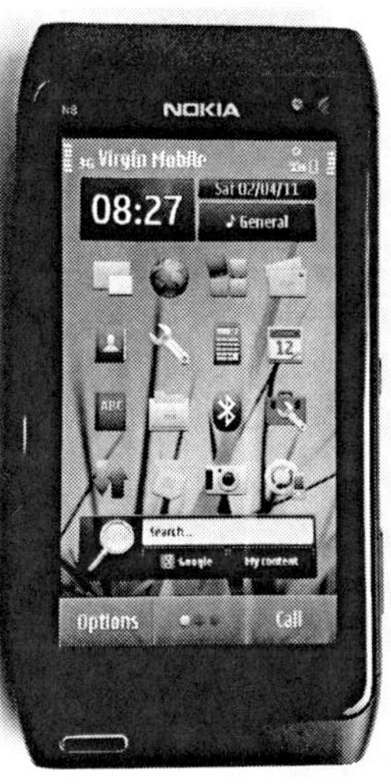

Nokia N8 running on Symbian^3 OS was Nokia's first camera phone to feature a 12 megapixel camera with Carl Zeiss Optics

Nokia stated that Maemo would be developed alongside Symbian. Maemo had since (Maemo "6" and beyond) merged with Intel's Moblin, and became MeeGo. MeeGo was later canceled and a development is now continued under name Tizen.

The Nokia N8 is the first device to function on the Symbian^3 mobile operating system.

Nokia revealed that the N8 will be the last device in its flagship N-series devices to ship with Symbian OS.[67][68]

Instead, Nokia will use Microsoft Windows Phone for its high-end flagship Lumia devices, and revealed the Nokia N9 will function on the MeeGo mobile operating system.

Alliance with Microsoft

On 11 February 2011, Nokia's CEO Stephen Elop, a former Microsoft employee, unveiled a new strategic alliance with Microsoft, and announced it would replace Symbian and MeeGo with Microsoft's Windows Phone operating system[69][70] except for mid-to-low-end devices, which would continue to run under Symbian. Nokia was also to invest into the Series 40 platform and release a single MeeGo product in 2011.[71]

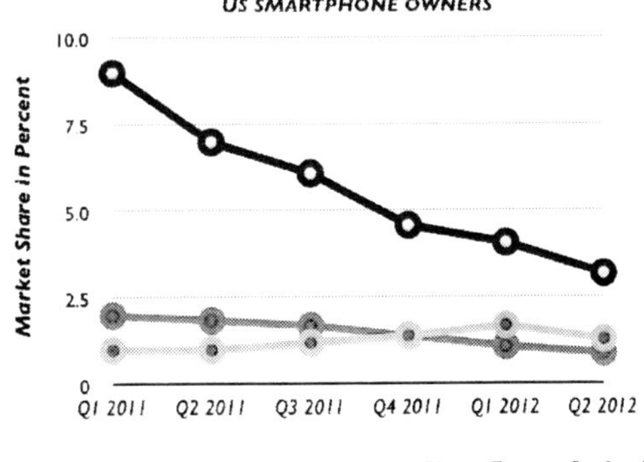

Market share of Symbian, Windows Mobile and Windows Phone 7 among US smartphone owners from Q1 2011 to Q2 2012 according to Nielsen Company.

As part of the restructuring plan, Nokia planned to reduce spending on research and development, instead customising and enhancing the software line for Windows Phone 7.[72] Nokia's "applications and content store" (Ovi) becomes integrated into the Windows Phone Marketplace, and Nokia Maps is at the heart of Microsoft's Bing and AdCenter. Microsoft provides developer tools to Nokia to replace the Qt framework, which is not supported by Windows Phone 7 devices.[73]

Symbian became described by Elop as a "franchise platform" with Nokia planning to sell 150 million Symbian devices after the alliance was set up. MeeGo emphasis was on longer-term exploration, with plans to ship "a MeeGo-related product" later in 2012. Microsoft's search engine, Bing was to become the search engine for all Nokia phones. Nokia also intended to get some level of customisation on WP7.[74]

After this announcement, Nokia's share price fell about 14%, its biggest drop since July 2009.[75]

As Nokia was the largest mobile phone manufacturer worldwide at the time,[76] it was suggested the alliance would make Microsoft's Windows Phone 7 a stronger contender against Android and iOS.[73] In June 2011 Nokia was overtaken by Apple as the world's biggest smartphone maker by volume.[77] In August 2011 Chris Weber, head of Nokia's subsidiary in the U.S., stated "*The reality is if we are not successful with Windows Phone, it doesn't matter what we do (elsewhere).*" He further added "*North America is a priority for Nokia (...) because it is a key market for Microsoft.*".[78] Additionally, while announcing an alliance with Groupon, Elop declared "The competition... is not with other device manufacturers, it's with Google."[79]

Nokia reported "well above 1 million" sales for its Lumia line up to January 26, 2012,[80][81] 2 million sales for the first quarter of 2012,[82] and 4 million for the second quarter of 2012.[83] In this quarter, Nokia only sold 600000 smartphones (Symbian and Windows Phone 7) in North America.[84] For comparison, 26 million iPhones and 105 million Android phones have been shipped worldwide in Q2 2012.[85]

European carriers have stated that Nokia Windows phones are not good enough to compete with Apple iPhone or Samsung Galaxy phones, that "they are overpriced for what is not an innovative product" and that "No one comes into the store and asks for a Windows phone".[86]

In June 2012, Nokia chairman Risto Siilasmaa told journalists that Nokia had a back-up plan in the eventuality that Windows Phone failed to be sufficiently successful in the market.[87][88]

Plant movements

Nokia opened its Komárom, Hungary mobile phone factory on 5 May 2000.[89]

In March 2007, Nokia signed a memorandum with Cluj County Council, Romania to open a new plant near the city in Jucu commune.[90][91][92] Moving the production from the Bochum, Germany factory to a low wage country created an uproar in Germany.[93][94] Nokia recently moved its North American Headquarters to Sunnyvale.

Reorganizations

In April 2003, the troubles of the networks equipment division caused the corporation to resort to similar streamlining practices on that side, including layoffs and organizational restructuring.[95] This diminished Nokia's public image in Finland,[96][97] and produced a number of court cases and an episode of a documentary television show critical of Nokia.[98]

On February 2006, Nokia and Sanyo announced a memorandum of understanding to create a joint venture addressing the CDMA handset business. But in June, they announced ending negotiations without agreement. Nokia also stated its decision to pull out of CDMA research and development, to continue CDMA business in selected markets.[99][100][101]

In June 2006, Jorma Ollila left his position as CEO to become the chairman of Royal Dutch Shell[102] and to give way for Olli-Pekka Kallasvuo.[103][104]

In May 2008, Nokia announced on their annual stockholder meeting that they want to shift to the Internet business as a whole. Nokia no longer wants to be seen as the telephone company. Google, Apple and Microsoft are not seen as natural competition for their new image but they are considered as major important players to deal with.[105]

In November 2008, Nokia announced it was ceasing mobile phone distribution in Japan.[106] Following early December, distribution of Nokia E71 is cancelled, both from NTT docomo and SoftBank Mobile. Nokia Japan retains global research & development programs, sourcing business, and an MVNO venture of Vertu luxury phones, using docomo's telecommunications network.

In February 2012, Nokia anonunced it was laying off 4000 employees to move manufacturing from Europe and Mexico to Asia.[107]

In March 2012, Nokia annonunced it was laying off 1000 employess from its Salo, Finland factory to focus on software.[108]

Acquisitions

For a more comprehensive list, see List of acquisitions by Nokia.

On 22 September 2003, Nokia acquired Sega.com, a branch of Sega which became the major basis to develop the Nokia N-Gage device.[109]

On 16 November 2005, Nokia and Intellisync Corporation, a provider of data and PIM synchronization software, signed a definitive agreement for Nokia to acquire Intellisync.[110] Nokia completed the acquisition on 10 February 2006.[111]

The Nokia E55 from the business segment of the Eseries range

On 19 June 2006, Nokia and Siemens AG announced the companies would merge their mobile and fixed-line phone network equipment businesses to create one of the world's largest network firms, Nokia Siemens Networks.[112] Each company has a 50% stake in the infrastructure company, and it is headquartered in Espoo, Finland. The companies predicted annual sales of €16 bn and cost savings of €1.5 bn a year by 2010. About 20,000 Nokia employees were transferred to this new company.

On 8 August 2006, Nokia and Loudeye Corp. announced that they had signed an agreement for Nokia to acquire online music distributor Loudeye Corporation for approximately US $60 million.[113] The company has been developing this into an online music service in the hope of using it to generate handset sales. The service, launched on 29 August 2007, is aimed to rival iTunes. Nokia completed the acquisition on 16 October 2006.[114]

In July 2007, Nokia acquired all assets of Twango, the comprehensive media sharing solution for organizing and sharing photos, videos and other personal media.[115][116]

In September 2007, Nokia announced its intention to acquire Enpocket, a supplier of mobile advertising technology and services.[117]

In October 2007, pending shareholder and regulatory approval, Nokia bought Navteq, a U.S.-based supplier of digital mapping data, for a price of $8.1 billion.[8][118] Nokia finalized the acquisition on 10 July 2008.[119]

In September, 2008, Nokia acquired OZ Communications, a privately held company with approximately 220 employees headquartered in Montreal, Canada.[120]

On 24 July 2009, Nokia announced that it will acquire certain assets of cellity, a privately owned mobile software company which employs 14 people in Hamburg, Germany.[121] The acquisition of cellity was completed on 5 August 2009.[122]

On 11 September 2009, Nokia announced the acquisition of "certain assets of Plum Ventures, Inc, a privately held company which employed approximately 10 people with main offices in Boston, Massachusetts. Plum will complement Nokia's Social Location services".[123]

On 28 March 2010, Nokia announced the acquisition of Novarra, the mobile web browser firm from Chicago. Terms of the deal were not disclosed. Novarra is a privately held company based in Chicago, IL and provider of a mobile browser and service platform and has more than 100 employees.[124]

On 10 April 2010, Nokia announced its acquisition of MetaCarta, whose technology was planned to be used in the area of local search, particularly involving location and other services. Financial details of acquisition were not disclosed.[125]

Nokia has acquired Smarterphone in 2012.[126] Also Nokia acquired Scalado in 2012.[127]

Financial difficulties and restructuring

Amid falling sales, Nokia posted a loss of 368 million euros for Q2 2011, while in Q2 2010 had still a profit of 227 million euros. On September 2011, Nokia has announced it will lose another 3,500 jobs worldwide, including the closure of its Cluj factory in Romania.[128]

On 8 February 2012 Nokia Corp. said to cut around 4,000 jobs at smartphone manufacturing plants in Europe by the end of 2012 to move assembly closer to component supplier in Asia. It plans to cut 2,300 of the 4,400 jobs in Hungary, 700 out of 1,000 jobs in Mexico, and 1,000 out of 1,700 factory jobs in Finland.[129]

On 14 June 2012, Nokia announced to cut 10,000 jobs globally by the end of 2013[130] and shut production and research sites in Finland, Germany and Canada inline with continues loss and the stock fell to the lowest since 1996. Today, Nokia's market value is below $10 billion.[131]

In total, according to actualized and planned laid-offs Nokia will have laid off 24,500 employees by the end of 2013. Nokia has already laid off 7,000 employees in the first stage: 4,000 staff and transferred also 3,000 to services firm Accenture. Nokia also closed its factory in Cluj, Romania that decreased the workforce by 2,000 employees, and restructured the Location & Commerce business unit that decreased the workforce by 1,200 employees. In February 2012, Nokia unveiled a plan to cut 4,000 more jobs at its plants in Finland, Hungary and Mexico as it moves smartphone assembly work to Asia. The most recent plan is to cut further 10,000 jobs globally by the end of 2013.[132] Nokia had 66,267 personnel in its Devices&Services, NAVTEQ and Corporate Common Functions units combined, this has been calculated by subtracting the personnel of Nokia Siemens Networks from the total personnel of Nokia Group based on the full year report of 2010.[133] Therefore, the personnel would decrease by approximately 36 percent by the end of 2013 when compared to the end of 2010 that best depicts the lay-offs that have resulted from the strategy change in February 2011 and competition in the central mobile phone business units recently.

On 18 June 2012 Moody's downgraded Nokia rating to junk.[134] Nokia CEO admitted on 28 June 2012 that company's inability to foresee rapid changes in mobile phone industry was one of the major reasons for the problems company was facing.[135]

Corporate affairs

Corporate structure

Divisions

Since 1 July 2010, Nokia comprises three business groups: **Mobile Solutions**, **Mobile Phones** and **Markets**.[136] The three units receive operational support from the **Corporate Development Office**, led by Kai Öistämö, which is also responsible for exploring corporate strategic and future growth opportunities.[136]

On 1 April 2007, Nokia's Networks business group was combined with Siemens's carrier-related operations for fixed and mobile networks to form Nokia Siemens Networks, jointly owned by Nokia and Siemens and consolidated by Nokia.[137]

Mobile Solutions

The Nokia N900, a Maemo 5 Linux based mobile Internet device and touchscreen smartphone from Nokia's Nseries portfolio.

Mobile Solutions is responsible for Nokia's portfolio of smartphones and mobile computers, including the more expensive multimedia and enterprise-class devices. The team is also responsible for a suite of internet services under the Ovi brand, with a strong focus on maps and navigation, music, messaging and media.[136] This unit is led by Anssi Vanjoki, along with Tero Ojanperä (for Services) and Alberto Torres (for MeeGo Computers).[136]

Alberto Torres has stepped down.

Mobile Phones

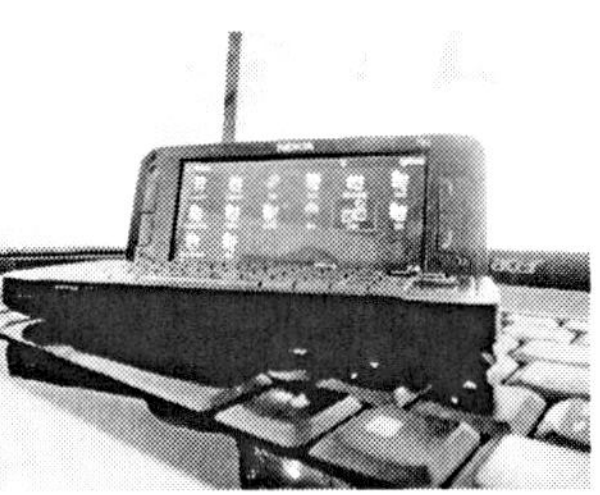

The Nokia E90, a Symbian smartphone from Nokia's Eseries portfolio.

Mobile Phones is responsible for Nokia's portfolio of affordable mobile phones, as well as a range of services that people can access with them, headed by Mary T. McDowell.[136] This unit provides the general public with mobile voice and data products across a range of devices, including high-volume, consumer oriented mobile phones. The devices are based on GSM/EDGE, 3G/W-CDMA and CDMA cellular technologies.

At the end of the year 2007, Nokia managed to sell almost 440 million mobile phones which accounted for 40% of all global mobile phones sales.[138] In 2011, Nokia's market share in the mobile phone market had dropped to 27% (417 million phones).[139]

Anssi Vanjoki resigned a few days before Nokia World 2010 and under new leadership team Jo Harlow will look into the affairs of Smartphones portfolio.

On 27 April 2011, The Register reported that Nokia was secretly developing a new operating system called Meltemi aiming at the low-end market. It was believed it would be replacing the S30 and S40 operating systems. Due to low-end market customers' demand of having smartphone features in their feature phone, the OS would have included some features exclusive to high-end smartphones. On 26 July 2012, it was announced that Nokia had abandoned the Meltemi project as a cost-cutting measure.

Markets

Markets is responsible for Nokia's supply chains, sales channels, brand and marketing functions of the company, and is responsible for delivering mobile solutions and mobile phones to the market. The unit is headed by Niklas Savander.[136]

Subsidiaries

Nokia has several subsidiaries, of which the two most significant as of 2009 are Nokia Siemens Networks and Navteq.[136] Other notable subsidiaries include, but are not limited to Vertu, a British-based manufacturer and retailer of luxury mobile phones; Qt Software, a Norwegian-based software company, and OZ Communications, a consumer e-mail and instant messaging provider.

Until 2008 Nokia was the major shareholder in Symbian Limited, a software development and licensing company that produced Symbian OS, a smartphone operating system used by Nokia and other manufacturers. In 2008 Nokia acquired Symbian Ltd and, along with a number of other companies, created the Symbian Foundation to distribute the Symbian platform royalty free and as open source.

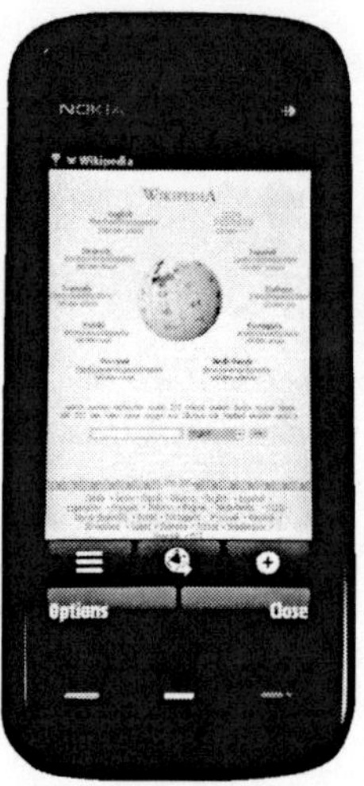

The Nokia 5800 XpressMusic, a touchscreen smartphone and portable entertainment device which emphasizes music and multimedia playback.

Nokia Siemens Networks

Nokia Siemens Networks (previously Nokia Networks) provides wireless and fixed network infrastructure, communications and networks service platforms, as well as professional services to operators and service providers.[136] Nokia Siemens Networks focuses in GSM, EDGE, 3G/W-CDMA and WiMAX radio access networks; core networks with increasing IP and multiaccess capabilities; and services.

On 19 June 2006 Nokia and Siemens AG announced the companies are to merge their mobile and fixed-line phone network equipment businesses to create one of the world's largest network firms, called Nokia Siemens Networks.[112] The Nokia Siemens Networks brand identity was subsequently launched at the 3GSM World Congress in Barcelona in February 2007.[140][141]

Navteq

Navteq is a Chicago, Illinois-based provider of digital map data and location-based content and services for automotive navigation systems, mobile navigation devices, Internet-based mapping applications, and government and business solutions.[136] Navteq was acquired by Nokia on 1 October 2007.[8] Navteq's map data is part of the Nokia Maps online service where users can download maps, use voice-guided navigation and other context-aware web services.[136] Nokia Maps is part of the Ovi brand of Nokia's Internet based online services.

Corporate governance

The control and management of Nokia is divided among the shareholders at a general meeting and the Nokia Leadership Team (left),[142] under the direction of the Board of Directors (right).[143] The Chairman and the rest of the Nokia Leadership Team members are appointed by the Board of Directors. Only the Chairman of the Nokia Leadership Team can belong to both, the Board of Directors and the Nokia Leadership Team. The Board of Directors' committees consist of the Audit Committee,[144] the Personnel Committee[145] and the Corporate Governance and Nomination Committee.[146][147]

The operations of the company are managed within the framework set by the Finnish Companies Act,[148] Nokia's Articles of Association[149] and Corporate Governance Guidelines,[150] and related Board of Directors adopted charters.

Nokia Leadership Team (as of April 2012) [142]
Stephen Elop (Chairman), b. 1963 President, CEO and Nokia Leadership Team Chairman of Nokia Corporation since 21 September 2010 Joined Nokia on 21 September 2010, Nokia Board member since 3 May 2011
Esko Aho, b. 1954 Executive Vice President, Corporate Relations and Responsibility Joined Nokia 2008, Nokia Leadership Team member since 2009 Former Prime Minister of Finland (1991–1995)
Marko Ahtisaari, b. 1969 Executive Vice President, Design Joined Nokia 2009, Nokia Leadership Team member since 1 February 2012
Jerri DeVard, b. 1958 Executive Vice President, Chief Marketing Officer Joined Nokia 2011, Nokia Leadership Team member since 1 January 2011
Colin Giles, b. 1963 Executive Vice President, Sales Joined Nokia 1992, Nokia Leadership Team member since 11 February 2011
Michael Halbherr, b. 1964 Executive Vice President, Location & Commerce Joined Nokia 2006, Nokia Leadership Team member since 1 July 2011
Jo Harlow, b. 1962 Executive Vice President, Smart Devices Joined Nokia 2003, Nokia Leadership Team member since 11 February 2011
Timo Ihamuotila, b. 1966 Executive Vice President, Chief Financial Officer With Nokia 1993–1996, rejoined 1999, Nokia Leadership Team member since 2007
Mary T. McDowell, b. 1964 Executive Vice President, Mobile Phones Joined Nokia 2004, Nokia Leadership Team member since 2004
Louise Pentland, b. 1972 Executive Vice President, Chief Legal Officer Joined Nokia 1998, Nokia Leadership Team member since 11 February 2011
Niklas Savander, b. 1962 Executive Vice President, Markets Joined Nokia 1997, Nokia Leadership Team member since 2006
Henry Tirri, b. 1956 Executive Vice President, Chief Technology Officer Joined Nokia 2004, Nokia Leadership Team member since 22 September 2011
Juha Äkräs, b. 1965 Executive Vice President, Human Resources Joined Nokia 1993, Nokia Leadership Team member since 2010
Dr. Kai Öistämö, b. 1964 Executive Vice President, Chief Development Officer Joined Nokia 1991, Nokia Leadership Team member since 2005

Board of Directors [143]
Risto Siilasmaa (Chairman), b. 1966 Board member since 2008, Chairman of the Board of Directors since 3 May 2012 Chairman of the Corporate Governance and Nomination Committee Founder and Chairman of F-Secure Corporation
Dame Marjorie Scardino (Vice Chairman), b. 1947 Board member since 2001, Vice Chairman since 2007 Member of the Corporate Governance and Nomination Committee, Member of the Personnel Committee Chief Executive Officer and member of the Board of Directors of Pearson PLC
Bruce Brown, b. 1958 Board member since 3 May 2012, Member of the Personnel Committee Chief Technology Officer of Procter & Gamble
Stephen Elop, b. 1963 Board member since 3 May 2011 President and CEO of Nokia Corporation, Chairman of the Nokia Leadership Team
Dr. Henning Kagermann, b. 1947 Board member since 2007, Chairman of the Personnel Committee, Member of the Corporate Governance and Nomination Committee Former CEO and Chairman of the Executive Board of SAP AG
Jouko Karvinen, b. 1957 Board member since 3 May 2011, Chairman of the Audit Committee, Member of the Corporate Governance and Nomination Committee CEO of Stora Enso Oyj
Helge Lund, b. 1962 Board member since 3 May 2011, Member of the Personnel Committee President and CEO of Statoil ASA
Isabel Marey-Semper, b. 1967 Board member since 2009, Member of the Audit Committee Director of Advanced Research of L'Oréal Group
Mårten Mickos, b. 1962 Board member since 3 May 2012 Chief Executive Officer of Eucalyptus Systems, Inc.
Elizabeth Nelson, b. 1960 Board member since 3 May 2012, Member of the Audit Committee Independent Corporate Advisor
Kari Stadigh, b. 1955 Board member since 3 May 2011, Member of the Personnel Committee Group CEO and President of Sampo plc

Former corporate officers

Chief Executive Officers		Chairmen of the Board of Directors [151]			
Björn Westerlund	1967–1977	Lauri J. Kivekäs	1967–1977	Simo Vuorilehto	1988–1990
Kari Kairamo	1977–1988	Björn Westerlund	1977–1979	Mika Tiivola	1990–1992
Simo Vuorilehto	1988–1992	Mika Tiivola	1979–1986	Casimir Ehrnrooth	1992–1999
Jorma Ollila	1992–2006	Kari Kairamo	1986–1988	Jorma Ollila	1999–2012
Olli-Pekka Kallasvuo	2006–2010				

International presence

In 2011 Nokia had 130,000 employees in 120 countries, sales in more than 150 countries, global annual revenue of over €38 billion, and operating loss of €1 billion.[4] It was the world's largest manufacturer of mobile phones in 2011, with global device market share of 23% in the second quarter.[76]

The Nokia Research Center, founded in 1986, is Nokia's industrial research unit consisting of about 500 researchers, engineers and scientists;[152][153] it has sites in seven countries: Finland, China, India, Kenya, Switzerland, the United Kingdom and the United States.[154] Besides its research centers, in 2001 Nokia founded (and owns) INdT – Nokia Institute of Technology, a R&D institute located in Brazil.[155] Nokia operates a total of 9 manufacturing facilities[11] located at Salo, Finland; Manaus, Brazil; Cluj, Romania; Beijing and Dongguan, China; Komárom, Hungary; Chennai, India; Reynosa, Mexico; and Changwon, South Korea.[90][156] Nokia's factory in Cluj was seized by the Romanian government in November 2011 to prevent a sale of the assets, after Nokia had accumulated a tax liability of US$ 10 million.[157] Nokia's industrial design department is headquartered in Soho in London, UK with significant satellite offices in Helsinki, Finland and Calabasas, California in the US.

Nokia is a public limited-liability company listed on the Helsinki, Frankfurt, and New York stock exchanges.[11] Nokia plays a very large role in the economy of Finland.[158][159] It is an important employer in Finland and several small companies have grown into large ones as its partners and subcontractors.[160] In 2009 Nokia contributed 1.6% to Finland's GDP, and accounted for about 16% of Finland's exports in 2006.[161]

In February 2012 Nokia announced that it was cutting 4,000 factory jobs in Finland, Hungary and Mexico (more than half of the 7,100 jobs at the three factories affected) and moving smartphone assembly to existing facilities in South Korea and China.[162]

Logos

Past

NOKIA
CONNECTING PEOPLE

Nokia introduced its "Connecting People" advertising slogan, coined by Ove Strandberg "HS Archives" (in Finnish). Helsingin Sanomat. 1 June 2003. . Retrieved 14 May 2008. and used since 1992. "NOKIA | Connecting Pople 1992 Vector Logo (AI EPS)". HDicon.com. . Retrieved 17 October 2010.This earlier version of the slogan used Times RomanTimes Roman SC (Small Caps) font.Pitkänen, Juhani (Nokia's Art Director) (3 September 2007). "Nokia Strategic Marketing, Brand Identity". Nokia Corporation. . Retrieved 14 May 2008.

Present

NOKIA
Connecting People

Nokia's current logo used since 2006, "NOKIA | Connecting Pople new Vector Logo (AI EPS)". HDicon.com. . Retrieved 17 October 2010. with the redesigned "Connecting People" slogan.This slogan originally used Nokia's proprietary 'Nokia Sans' font, designed by Erik Spiekermann. "Erik Spiekermann – Furniture, Designs & Home Decor". Design Within Reach. . Retrieved 7 January 2010. This was replaced in 2011 with the 'Nokia Pure' font designed by Dalton Maag. "Our New Typeface". Nokia Little Blog of Branding. . Retrieved 3 April 2012.

Stock

Nokia, a public limited liability company, is the oldest company listed under the same name on the Helsinki Stock Exchange (since 1915).[24] Nokia's shares are also listed on the Frankfurt Stock Exchange (since 1988) and New York Stock Exchange (since 1994).[11][24]

In 2007, Nokia was valued at €110 billion; as of May 2012, it was valued at €14.8 billion. [169]

For fiscal Q2 2011 ending in June 2011, Nokia reported a net loss of €492 million, despite a €430 million payment from Apple. Nokia cited decline in its mobile phone business as the primary cause of the loss.[170]

In Q1 2012 results were bleak. Nokia lost €1.34 billion. Revenue is down almost a third from a year ago.[171] By May 2012, Nokia share price had fallen 37.5 percent since the beginning of the year, and was down 61 percent in the last year.[172][173]

Corporate culture

Nokia's official corporate culture manifesto, *The Nokia Way*, emphasises the speed and flexibility of decision-making in a flat, networked organization, although the corporation's size necessarily imposes a certain amount of bureaucracy.[174]

The Nokia House, Nokia's head office in Keilaniemi, Espoo, Finland.

The official business language of Nokia is English. All documentation is written in English, and is used in official intra-company spoken communication and e-mail.

Until May 2007, the *Nokia Values* were Customer Satisfaction, Respect, Achievement, and Renewal. In May 2007, Nokia redefined its values after initiating a series of discussions worldwide as to what the new values of the company should be. Based on the employee suggestions, the new values were defined as: Engaging You, Achieving Together, Passion for Innovation and Very Human.[174]

Online services

.mobi and the Mobile Web

Nokia was the first proponent of a Top Level Domain (TLD) specifically for the Mobile Web and, as a result, was instrumental in the launch of the .mobi domain name extension in September 2006 as an official backer.[175][176] Since then, Nokia has launched the largest mobile portal, Nokia.mobi, which receives over 100 million visits a month.[177] It followed that with the launch of a mobile Ad Service to cater to the growing demand for mobile advertisement.[178]

Ovi

Ovi, announced on 29 August 2007, is the name for Nokia's "umbrella concept" Internet services.[179] Centered on Ovi.com, it is marketed as a "personal dashboard" where users can share photos with friends, download music, maps and games directly to their phones and access third-party services like Yahoo's Flickr photo site. It has some significance in that Nokia is moving deeper into the world of Internet services, where head-on competition with Microsoft, Google and Apple is inevitable.[180]

The services offered through Ovi include the Ovi Store (Nokia's application store), the Nokia Music Store, Nokia Maps, Ovi Mail, the N-Gage mobile gaming platform available for several S60 smartphones, Ovi Share, Ovi Files, and Contacts and Calendar.[181] The Ovi Store, the Ovi application store was launched in May 2009.[182] Prior to opening the Ovi Store, Nokia integrated its software Download! store, the stripped-down MOSH repository and the widget service WidSets into it.[183]

On 23 March 2010, Nokia announced launch of its online magazine called the *Nokia Ovi*. The 44-page magazine contains articles on products by Nokia, what Ovi stands for, tips and tricks on the usage of Nokia mini laptop Booklet 3G, latest reviews of mobile applications, news about the mobile maker's services and apps such as Ovi maps, files and mail. Users can download the magazine as a PDF or view it online from the Nokia website.[184]

My Nokia

Nokia offers a free personalised service to Nokia owners called My Nokia (located at my.nokia.com).[185] Registered My Nokia users can get free services as follows:

- Tips & tricks alerts through web, e-mail and also mobile text message.
- My Nokia Backup: A free online backup service for mobile contacts, calendar logs and also various other files. This service needs GPRS connection.
- Ringtones, wallpapers, screensavers, games and other things can be downloaded free of cost.

Comes With Music

In 2007 Nokia set up their "Nokia Comes With Music" service, in partnership with Universal Music Group International, Sony BMG, Warner Music Group, EMI, and hundreds of independent labels and music aggregators, to allow 12, 18, or 24 months of unlimited free-of-charge music downloads with the purchase of a Nokia Comes With Music edition phone. Files could be downloaded on mobile devices or personal computers, and kept permanently.[61]

In January 2011 Nokia withdrew this program in 27 countries, due to its failure to gain traction with customers or mobile network operators; existing subscribers could continue to download until their contracts ended. The service continued to be offered in China, India, Indonesia, Brazil, Turkey and South Africa where take-up had been better.[186]

Nokia Messaging

On 13 August 2008 Nokia launched a beta release of "Nokia Email service", a push e-mail service, since incorporated into Nokia Messaging.[187]

Nokia Messaging operates as a centralised, hosted service that acts as a proxy between the Nokia Messaging client and the user's e-mail server. The phone does not connect directly to the e-mail server, but instead sends e-mail credentials to Nokia's servers.[188] IMAP is used as the protocol to transfer emails between the client and the server.

Controversies

NSN's provision of intercept capability to Iran

In 2008, Nokia Siemens Networks, a joint venture between Nokia and Siemens AG, reportedly provided Iran's monopoly telecom company with technology that allowed it to intercept the Internet communications of its citizens to an unprecedented degree.[189] The technology reportedly allowed it to use deep packet inspection to read and even change the content of everything from "e-mails and Internet phone calls to images and messages on social-networking sites such as Facebook and Twitter". The technology "enables authorities to not only block communication but to monitor it to gather information about individuals, as well as alter it for disinformation purposes," expert insiders told *The Wall Street Journal*. During the post-election protests in Iran in June 2009, Iran's Internet access was reported to have slowed to less than a tenth of its normal speeds, and experts suspected this was due to the use of the interception technology.[190]

The joint venture company, Nokia Siemens Networks, asserted in a press release that it provided Iran only with a 'lawful intercept capability' "solely for monitoring of local voice calls". "Nokia Siemens Networks has not provided any deep packet inspection, web censorship or Internet filtering capability to Iran," it said.[191]

In July 2009, Nokia began to experience a boycott of their products and services in Iran. The boycott was led by consumers sympathetic to the post-election protest movement and targeted at those companies deemed to be collaborating with the Islamic regime. Demand for handsets fell and users began shunning SMS messaging.[192]

Lex Nokia

In 2009, Nokia heavily supported the passing of a law in Finland that allows companies to monitor their employees' electronic communications in cases of suspected information leaking.[193] Contrary to rumors, Nokia denied that the company would have considered moving its head office out of Finland if laws on electronic surveillance were not changed.[194] The law was enacted, but with strict requirements for implementation of its provisions. As of 2010, the law has become a dead letter; no corporation has implemented it. The Finnish media dubbed the name *Lex Nokia* for this law, named after the Finnish copyright law (the so-called *Lex Karpela*) a few years back.

Nokia–Apple patent dispute

In October 2009, Nokia filed a lawsuit against Apple Inc. in the U.S. District Court of Delaware citing Apple infringed on 10 of its patents related to wireless communication including data transfer.[195] Apple was quick to respond with a countersuit filed in December 2009 accusing Nokia of 11 patent infringements. Apple's General Counsel, Bruce Sewell went a step further by stating, "Other companies must compete with us by inventing their own technologies, not just by stealing ours." This resulted in an ugly spat between the two telecom majors with Nokia filing another suit, this time with the U.S. International Trade Commission (ITC), alleging Apple of infringing its patents in "virtually all of its mobile phones, portable music players, and computers."[196] Nokia went on to ask the court to bar all U.S. imports of the Apple products including the iPhone, Mac and the iPod. Apple countersued by filing a complaint with the ITC in January 2010, the details of which are yet to be confirmed.[195]

In June 2011, Apple settled with Nokia and agreed to an estimated one time payment of $600 million and royalties to Nokia.[197] The two companies also agreed on a cross-licensing patents for some of their patented technologies.[198][199]

Environmental record

Electronic products such as cell phones impact the environment both during production and after their useful life when they are discarded and turned into electronic waste. Nokia is listed in Greenpeace's Guide to Greener Electronics that scores leading electronics manufacturers according to their policies on sustainability, climate and energy and how green their products are. In November 2011 Nokia ranked 3rd out of 15 listed electronics companies, falling two places due to its weaker performance on the Energy criteria and scoring 4.9/10.[200]

All of Nokia's mobile phones are free of toxic polyvinyl chloride (PVC) since the end of 2005 and all new models of mobile phones and accessories launched in 2010 are on track to be free of brominated compounds, chlorinated flame retardants and antimony trioxide.[200]

Nokia's voluntary take-back programme to recycle old mobile phones spans 84 countries with almost 5,000 collection points.[201] However, the recycling rate of Nokia phones was only 3–5% in 2008, according to a global consumer survey released by Nokia.[202] The majority of old mobile phones are simply lying in drawers at home and very few old devices, about 4%, are being thrown into landfill and not recycled.[202]

All of Nokia's new models of chargers meet or exceed the Energy Star requirements.[203] Nokia aims to reduce its carbon dioxide emissions by at least 18 percent in 2010 from a baseline year of 2006 and cover 50 percent of its energy needs through renewable energy sources.[204] Greenpeace is challenging the company to use its influence at the political level as number 85 on the Fortune 500 to advocate for climate legislation and call for global greenhouse gas emissions to peak by 2015.[205]

Nokia is researching the use of recycled plastics in its products, which are currently used only in packaging but not yet in mobile phones.[206]

Since 2001, Nokia has provided eco declarations of all its products and since May 2010 provides Eco profiles for all its new products.[207] In an effort to further reduce their environmental impact in the future, Nokia released a new phone concept, Remade, in February 2008.[208] The phone has been constructed of solely recyclable materials.[208] The outer part of the phone is made from recycled materials such as aluminium cans, plastic bottles, and used car tires.[209] The screen is constructed of recycled glass, and the hinges have been created from rubber tires. The interior of the phone is entirely constructed with refurbished phone parts, and there is a feature that encourages energy saving habits by reducing the backlight to the ideal level, which then allows the battery to last longer without frequent charges.

Research cooperation with universities

Nokia is actively exploring and engaging in open innovation through selective research collaborations with major universities and institutions by sharing resources and leveraging ideas. Major research collaboration is with Tampere University of Technology based in Finland. Current collaborations include:[210]

- Aalto University School of Science and Technology, Finland
- École Polytechnique Fédérale de Lausanne, Switzerland
- ETH Zurich, Switzerland
- Massachusetts Institute of Technology, United States
- Stanford University, United States
- Tampere University of Technology, Finland
- Tsinghua University, China
- University of California, Berkeley, United States
- University of Cambridge, United Kingdom
- University of Southern California, United States

Awards and recognition

The Brand Trust Report [211] published by Trust Research Advisory has ranked Nokia in the 1st position among the brands in India.

References

[1] http://www.nasdaqomxnordic.com/shares/shareinformation?Instrument=HEX24311

[2] http://www.nyse.com/about/listed/quickquote.html?ticker=nok

[3] http://www.boerse-frankfurt.de/en/equities/search/result?name_isin_wkn=NOA3

[4] "Annual Results 2011" (http://www.results.nokia.com/results/Nokia_results2011Q4e.pdf) (PDF). Nokia Corporation. 26 January 2012. . Retrieved 2 March 2012.

[5] http://www.results.nokia.com/results/Nokia_results2012Q1e.pdf

[6] http://www.nokia.com/

[7] "Nokia in brief (2007)" (http://www.nokia.com/NOKIA_COM_1/About_Nokia/Sidebars_new_concept/Nokia_in_brief/InBriefJuly08.pdf) (PDF). Nokia Corporation. March 2008. . Retrieved 14 May 2008.

[8] "Nokia to acquire NAVTEQ" (http://www.nokia.com/A4136001?newsid=1157198) (Press release). Nokia Corporation. 1 October 2007. . Retrieved 22 March 2009.

[9] "Company" (http://www.nokiasiemensnetworks.com/global/AboutUs/Company/?languagecode=en). Nokia Siemens Networks. . Retrieved 14 July 2009.

[10] "Samsung overtakes Nokia in mobile phone shipments" (http://www.bbc.co.uk/news/business-17865117). BBC News. 27 April 2012. . Retrieved 27 May 2012.

[11] "Nokia – FAQ" (http://www.nokia.com/about-nokia/company/faq). Nokia Corporation. . Retrieved 16 March 2009.

[12] "*Global 500* 2011" (http://money.cnn.com/magazines/fortune/global500/2011/full_list/101_200.html). Fortune. 2011. . Retrieved 12 November 2011.

[13] https://www.google.com/finance?client=ob&q=NYSE:NOK

[14] http://247wallst.com/2012/05/14/almost-nothing-can-save-nokia-except-maybe-widespread-poverty/#ixzz1v8jU0K5J

[15] "Nokia's first Windows Phone 7 handset: Is it enough?" (http://www.bbc.co.uk/news/technology-15460569). BBC News. 26 October 2011. . Retrieved 27 October 2011.

[16] "Nokia – Nokia's first century – Story of Nokia" (http://www.nokia.com/about-nokia/company/story-of-nokia/nokias-first-century). Nokia Corporation. . Retrieved 16 March 2009.

[17] "Nokia – The birth of Nokia – Nokia's first century – Story of Nokia" (http://www.nokia.com/about-nokia/company/story-of-nokia/nokias-first-century/the-birth-of-nokia). Nokia Corporation. . Retrieved 16 March 2009.

[18] Helen, Tapio. "Idestam, Fredrik (1838–1916)" (http://www.kansallisbiografia.fi/english/?id=4296). Biographical Centre of the Finnish Literature Society. . Retrieved 22 March 2009.

[19] "Nokian Footwear: History" (http://www.nokianfootwear.fi/eng/our_story/). Nokian Footwear. . Retrieved 21 March 2009.

[20] Palo-oja, Ritva; Willberg, Leena (1998) (in Finnish). *Kumi – Kumin ja Suomen kumiteollisuuden historia*. Tampere, Finland: Tampere Museums. pp. 43–53. ISBN 978-951-609-065-1.

[21] "Finnish Cable Factory – Brief History" (http://web.archive.org/web/20070705233815/http://www.kaapelitehdas.fi/php/image.php?id=4856) (PDF). Kaapelitehdas.fi (http://www.kaapelitehdas.fi). Archived from the original (http://www.kaapelitehdas.fi/php/image.php?id=4856) on 5 July 2007. . Retrieved 16 March 2009.

[22] "Nokia – Verner Weckman – Nokia's first century – Story of Nokia" (http://www.nokia.com/about-nokia/company/story-of-nokia/nokias-first-century/verner-weckman). Nokia Corporation. . Retrieved 20 March 2009.

[23] "Nokia – The merger – Nokia's first century – Story of Nokia" (http://www.nokia.com/about-nokia/company/story-of-nokia/nokias-first-century/the-merger). Nokia Corporation. . Retrieved 16 March 2009.

[24] "Nokia – Towards Telecommunications" (http://www.nokia.com/NOKIA_COM_1/About_Nokia/Sidebars_new_concept/Broschures/TowardsTelecomms.pdf) (PDF). Nokia Corporation. August 2000. . Retrieved 5 June 2008.

[25] "Nokia – First electronic dept – Nokia's first century – Story of Nokia" (http://www.nokia.com/about-nokia/company/story-of-nokia/nokias-first-century/first-electronic-dept). Nokia Corporation. . Retrieved 16 March 2009.

[26] "Nokia – Jorma Ollila – Mobile revolution – Story of Nokia" (http://www.nokia.com/about-nokia/company/story-of-nokia/mobile-revolution/jorma-ollila). Nokia Corporation. . Retrieved 21 March 2009.

[27] "History in brief" (http://www.nokiantyres.com/history-in-brief). Nokian Tyres. . Retrieved 22 March 2009.

[28] Kaituri, Tommi (2000). "Automaattisten puhelinkeskusten historia" (http://www.cs.helsinki.fi/u/kerola/tkhist/k2000/alustukset/puhelinkeskukset/) (in Finnish). . Retrieved 21 March 2009.

[29] Palmberg, Christopher; Martikainen, Olli (23 May 2003). "Overcoming a Technological Discontinuity – The Case of the Finnish Telecom Industry and the GSM" (http://www.etla.fi/files/677_dp855.pdf) (PDF). The Research Institute of the Finnish Economy. . Retrieved 14 June 2009.

[30] "Puolustusvoimat: Kalustoesittely – Sanomalaitejärjestelmä" (http://www.mil.fi/maavoimat/kalustoesittely/index.dsp?level=81) (in Finnish). The Finnish Defence Forces. 15 June 2005. . Retrieved 14 May 2008.

[31] "The Finnish Defence Forces: Presentation of equipment: Message device" (http://www.mil.fi/maavoimat/kalustoesittely/00030_en.dsp). The Finnish Defence Forces. . Retrieved 14 June 2009.

[32] "Nokia 1100 phone offers reliable and affordable mobile communications for new growth markets" (http://press.nokia.com/PR/200308/915317_5.html) (Press release). Nokia Corporation. 27 August 2003. . Retrieved 26 May 2009.

[33] "Nokia – Mobira Cityman – The move to mobile – Story of Nokia" (http://www.nokia.com/about-nokia/company/story-of-nokia/the-move-to-mobile/mobira-cityman). Nokia Corporation. . Retrieved 14 May 2008.

[34] Juutilainen, Matti. "Siirtyvä tietoliikenne, luennot 7–8: Matkapuhelinverkot" (http://www.it.lut.fi/kurssit/06-07/Ti5312600/luentokalvot/luento07-08.pdf) (in Finnish) (PDF). Lappeenranta University of Technology. . Retrieved 22 March 2009.

[35] "Nokia – Mobile era begins – The move to mobile – Story of Nokia" (http://www.nokia.com/about-nokia/company/story-of-nokia/the-move-to-mobile/mobile-era-begins). Nokia Corporation. . Retrieved 20 March 2009.

[36] Karttunen, Anu (2 May 2003). "Tähdet syöksyvät, Benefon" (http://www.talouselama.fi/sijoittaminen/article165594.ece) (in Finnish). *Talouselämä* (Talentum Oyj). . Retrieved 28 July 2009.

[37] "Nokia´s Pioneering GSM Research and Development to be Awarded by Eduard Rhein Foundation" (http://press.nokia.com/1997/10/17/nokiaÂ´s-pioneering-gsm-research-and-development-to-be-awarded-by-eduard-rhein-foundation) (Press release). Nokia Corporation. 17 October 1997. . Retrieved 7 April 2012.

[38] "Global Mobile Communication is 20 years old" (http://gsmworld.com/newsroom/press-releases/2070.htm) (Press release). GSM Association. 6 September 2007. . Retrieved 23 March 2009.

[39] "Happy 20th birthday, GSM" (http://news.zdnet.co.uk/leader/0,1000002982,39289154,00.htm). *ZDNet.co.uk* (CBS Interactive). 7 September 2007. . Retrieved 23 March 2009.

[40] "Nokia – First GSM call – The move to mobile – Story of Nokia" (http://www.nokia.com/about-nokia/company/story-of-nokia/the-move-to-mobile/first-gsm-call). Nokia Corporation. . Retrieved 20 March 2009.

[41] Smith, Tony (9 November 2007). "15 years ago: the first mass-produced GSM phone" (http://www.reghardware.co.uk/2007/11/09/ft_nokia_1011/). *Register Hardware*. Situation Publishing Ltd. . Retrieved 23 March 2009.

[42] "Nokia – Nokia Tune – Mobile revolution – Story of Nokia" (http://www.nokia.com/about-nokia/company/story-of-nokia/mobile-revolution/nokia-tune). Nokia Corporation. . Retrieved 23 March 2009.

[43] "3 Billion GSM Connections On The Mobile Planet – Reports The GSMA" (http://www.gsmworld.com/newsroom/press-releases/2008/1108.htm). GSM Association. 16 April 2008. . Retrieved 21 March 2009.

[44] "Nokia MikroMikko 1" (http://www.old-computers.com/museum/computer.asp?st=1&c=630). Old-Computers.com. . Retrieved 14 May 2008.

[45] "Net – Fujitsun asiakaslehti, Net-lehden historia: 1980-luku" (http://www.fujitsuservices.fi/historia/net/1980.htm) (in Finnish). Fujitsu Services Oy, Finland. . Retrieved 22 March 2009.

[46] "Historia: 1991–1999" (http://www.fujitsu.com/fi/about/history/1991/) (in Finnish). Fujitsu Services Oy, Finland. . Retrieved 22 March 2009.

[47] Hietanen, Juha (28 February 2000). "Closure of Fujitsu Siemens plant – a repeat of Renault Vilvoorde?" (http://www.eiro.eurofound.eu.int/2000/02/feature/fi0002136f.html). EIRO, European Industrial Relations Observatory on-line. . Retrieved 14 May 2008.

[48] Hietanen, Juha (28 February 2000). "Fujitsu Siemens tehdas suljetaan – toistuiko Renault Vilvoord?" (http://www.eurofound.europa.eu/eiro/2000/02/word/fi0002136ffi.doc) (in Finnish) (DOC). EIRO, European Industrial Relations Observatory on-line. . Retrieved 14 May 2008.

[49] "ViewSonic Corporation Acquires Nokia Display Products' Branded Business" (http://press.nokia.com/PR/200001/775025_5.html) (Press release). Nokia Corporation. 17 January 2000. . Retrieved 22 March 2009.

[50] "Nokia Booklet 3G brings all day mobility to the PC world" (http://www.nokia.com/press/press-releases/showpressrelease?newsid=1336683) (Press release). Nokia Corporation. 24 August 2009. . Retrieved 26 August 2009.

[51] Pietilä, Antti-Pekka (27 September 2000). "Kari Kairamon nousu ja tuho" (http://www.taloussanomat.fi/arkisto/2000/09/27/kari-kairamon-nousu-ja-tuho/200026243/12) (in Finnish). *Taloussanomat*. . Retrieved 21 March 2009.

[52] "Finland: How bad policies turned bad luck into a recession" (http://www.cepr.org/PRESS/EP29 finland.htm). Centre for Economic Policy Research. . Retrieved 5 April 2009.

[53] Häikiö, Martti; translated by Hackston, David (2001) (in Finnish). *Nokia Oyj:n historia 1–3 (A history of Nokia plc 1–3)* (http://www.finlit.fi/booksfromfinland/bff/102/nokia.htm). Helsinki: Edita. ISBN 951-37-3467-6. . Retrieved 21 March 2008.

[54] "Nokia – Leading the world – Mobile revolution – Story of Nokia" (http://www.nokia.com/about-nokia/company/story-of-nokia/mobile-revolution/leading-the-world). Nokia Corporation. . Retrieved 21 March 2009.

[55] Reinhardt, Andy (3 August 2006). "Nokia's Magnificent Mobile-Phone Manufacturing Machine" (http://www.businessweek.com/globalbiz/content/aug2006/gb20060803_618811.htm). *BusinessWeek Online Europe*. . Retrieved 21 March 2009.

[56] Professor Voomann, Thomas E.; Cordon, Carlos (1998). "Nokia Mobile Phones: Supply Line Management" (http://stuff.mit.edu/afs/athena/course/15/15.795/Nokia Supply Chain Case Study.pdf) (PDF). Lausanne, Switzerland: IMD – International Institute for Management Development. . Retrieved 21 March 2009.

[57] Ewing, Jack (30 July 2007). "Why Nokia Is Leaving Moto in the Dust" (http://www.businessweek.com/magazine/content/07_31/b4044050.htm). *BusinessWeek Online*. . Retrieved 21 March 2009.

[58] Lin, Porter; Khan, Raedeep; Piekute, Vaida; Luhtasela, Jussi; Fang, Debby (1 December 2005). "Supply Chain Management Case Nokia" (http://imba.nccu.edu.tw/OIP/EXchange/Docs/F04/mis/final/group6/SCM in Nokia - Written report-V1.0.pdf) (PDF). IMBA, College of Commerce, National Chengchi University. . Retrieved 21 March 2009.

[59] Virki, Tarmo (5 March 2007). "Nokia's cheap phone tops electronics chart" (http://www.reuters.com/article/technologyNews/idUSL0262945620070503). Reuters. . Retrieved 14 May 2008.

[60] Saylor, Michael (2012). *The Mobile Wave: How Mobile Intelligence Will Change Everything*. Perseus Books/Vanguard Press. p. 304.

[61] "Nokia World 2007: Nokia outlines its vision of Internet evolution and commitment to environmental sustainability" (http://www.nokia.com/A4136001?newsid=1172937) (Press release). Nokia Corporation. 4 December 2007. . Retrieved 14 May 2008.

[62] "Nokia Productions and Spike Lee premiere the world's first social film" (http://www.nokia.com/press/press-releases/showpressrelease?newsid=1259528) (Press release). Nokia Corporation. 14 October 2008. . Retrieved 12 June 2009.

[63] "Nokia seizes social internet and amplifies music experience" (http://www.nokia.com/press/press-releases/showpressrelease?newsid=1338896) (Press release). Nokia Corporation. 2 September 2009. . Retrieved 12 October 2009.

[64] "Nokia 7705 Twist launched Stateside on Verizon (photo gallery)" (http://conversations.nokia.com/2009/09/10/nokia-7705-twist-launched-stateside-on-verizon-photo-gallery/). Nokia Corporation. 10 September 2009. . Retrieved 11 October 2009.

[65] "Nokia launches two new handsets under 'Asha' range" (http://economictimes.indiatimes.com/tech/hardware/nokia-launches-two-new-handsets-under-asha-range/articleshow/15421139.cms). 09-08-2012. .

[66] "Home of the Maemo community" (http://maemo.org/). maemo.org. . Retrieved 12 November 2011.

[67] Paul, Ryan (25 June 2010). "Nokia picks MeeGo Linux, not Symbian, for flagship phones" (http://arstechnica.com/open-source/news/2010/06/nokia-to-use-meego-linux-and-not-symbian-for-flagship-phones.ars). Ars Technica. . Retrieved 12 November 2011.

[68] Sherwood, James (13 August 2009). "Nokia exec denies Symbian Maemo swap claim" (http://www.reghardware.co.uk/2009/08/13/nokia_denies_maemo/). reghardware. . Retrieved 12 November 2011.

[69] "Nokia announces strategic partnership with Microsoft, will use WP7 as primary OS" (http://www.techit.in/2011/02/nokia-announces-strategic-partnership-with-microsoft-will-use-wp7-as-primary-os/). TechIt.in. .

[70] "Missed the historic Nokia+Microsoft event today? See it here!" (http://www.techit.in/2011/02/missed-the-historic-nokiamicrosoft-event-today-see-it-here/). TechIt.in. .

[71] "Nokia and Microsoft form partnership" (http://www.bbc.co.uk/news/business-12427680). BBC. 11 February 2011. . Retrieved 12 February 2011.

[72] "RIP: Symbian" (http://www.engadget.com/2011/02/11/rip-symbian/). Engadget. .
[73] "Capitulation: Nokia adopts Windows Phone 7" (http://arstechnica.com/gadgets/news/2011/02/nokia-adopts-windows-phone-7-as-primary-platform.ars). ArsTechnica. 11 February 2011. . Retrieved 12 February 2011.
[74] "Nokia will be able to customize 'everything' in Windows Phone 7, but likely won't" (http://www.engadget.com/2011/02/11/nokia-will-be-able-to-customize-everything-in-windows-phone-7). Engadget. .
[75] ben-Aaron, Diana (11 February 2011). "Nokia Falls Most Since July 2009 After Microsoft Deal" (http://www.bloomberg.com/news/2011-02-11/nokia-joins-forces-with-microsoft-to-challenge-dominance-of-apple-google.html). Bloomberg. .
[76] "Gartner Says Sales of Mobile Devices in Second Quarter of 2011 Grew 16.5 Percent Year-on-Year; Smartphone Sales Grew 74 Percent" (http://www.gartner.com/it/page.jsp?id=1764714) (Press release). Gartner. 11 August 2011. . Retrieved 29 September 2011.
[77] Ward, Andrew (21 July 2011). "Apple overtakes Nokia in smartphone stakes" (http://www.ft.com/cms/s/0/4d7fd1e2-b38e-11e0-b56c-00144feabdc0.html#axzz1SlVal4Cs). *Financial Times*. . Retrieved 21 July 2011.
[78] Fried, Ina (9 August 2011). "Nokia to Exit Symbian, Low-End Phone Businesses in North America" (http://allthingsd.com/20110809/exclusive-nokia-to-exit-symbian-low-end-phone-businesses-in-north-america/). All Things Digital. . Retrieved 9 August 2011.
[79] "Nokia's Lumia phones to show Groupon offers on maps" (http://www.bbc.com/news/technology-19093769). BBC. .
[80] http://press.nokia.com/2012/01/26/nokia-q4-2011-net-sales-eur-10-0-billion-non-ifrs-eps-eur-0-06-reported-eps-eur-0-29-nokia-2011-net-sales-eur-38-7-billion-non-ifrs-eps-eur-0-29-reported-eps-e
[81] http://www.intomobile.com/2012/01/26/nokia-q4-2011-financial-results-over-1-million-lumia-windows-phones-sold-date/
[82] Cooper, Daniel. "Nokia: Two million Lumia phones sold in Q1 but profits still falling" (http://www.engadget.com/2012/04/11/nokia-2012-q1-forecast/). *Engadget*. AOL. . Retrieved June 2, 2012.
[83] "Nokia ships 4 million Lumias in Q2 2012, over 7 million to date" (http://www.theverge.com/2012/7/19/3169153/nokia-lumia-sales-4-million-q2-2012). .
[84] "Nokia's Fight for AT&T Shelf Space Shows Hurdle for Lumia" (http://www.bloomberg.com/news/2012-08-05/nokia-fighting-for-at-t-shelf-space-shows-hurdle-for-next-lumia.html). .
[85] "Android smartphone share quadruples iOS in Q2" (http://news.cnet.com/8301-1035_3-57488926-94/android-smartphone-share-quadruples-ios-in-q2/). .
[86] Mobile operators unconvinced by Nokia's revival bid | Reuters (http://uk.reuters.com/article/2012/04/17/uk-nokia-telcos-idUKBRE83G08Z20120417)
[87] Chris Smith (30 June 2012). "Nokia promises back-up plan if Windows Phone fails" (http://www.techradar.com/news/phone-and-communications/mobile-phones/nokia-promises-back-up-plan-if-windows-phone-fails-1087600). Techradar. . Retrieved 6 July 2012.
[88] "Nokia's Siilasmaa: Goal to regain competitiveness" (http://yle.fi/uutiset/nokias_siilasmaa_goal_to_regain_competitiveness/6199219). YLS Uutiset. 28 June 2012. . Retrieved 6 July 2012.
[89] "Hungarian and Finnish Prime Ministers Inaugurate Nokia's "Factory of the Future" in Komárom" (http://press.nokia.com/PR/200005/780293_5.html) (Press release). Nokia Corporation. 5 May 2000. . Retrieved 22 March 2009.
[90] "Nokia to set up a new mobile device factory in Romania" (http://www.nokia.com/A4136001?newsid=1114420) (Press release). Nokia Corporation. 26 March 2007. . Retrieved 14 May 2008.
[91] "Nokia to open cell phone plant near Cluj" (http://web.archive.org/web/20080507194303/http://www.boston.com/news/world/europe/articles/2007/03/22/nokia_to_open_cell_phone_plant_near_cluj/). Associated Press. Boston.com. 22 March 2007. Archived from the original (http://www.boston.com/news/world/europe/articles/2007/03/22/nokia_to_open_cell_phone_plant_near_cluj/) on 7 May 2008. . Retrieved 14 May 2008.
[92] "Nokia to build mobile phone plant in Romania" (http://www.hs.fi/english/article/Nokia+to+build+mobile+phone+plant+in+Romania/1135226144930). *Helsingin Sanomat*. 27 March 2007. . Retrieved 14 May 2008.
[93] "German Politicians Return Cell Phones Amid Nokia Boycott Calls" (http://www.dw-world.de/dw/article/0,2144,3076534,00.html). *Deutsche Welle*. 18 January 2008. . Retrieved 22 March 2009.
[94] "German State Demands €60 Million from Nokia" (http://www.spiegel.de/international/business/0,1518,540699,00.html). *Der Spiegel*. 11 March 2008. . Retrieved 22 March 2009.
[95] "Nokia Networks takes strong measures to reduce costs, improve profitability and strengthen leadership position" (http://press.nokia.com/PR/200304/898905_5.html) (Press release). Nokia Corporation. 10 April 2003. . Retrieved 14 May 2008.
[96] "Nokia Networks to shed 1,800 jobs worldwide; majority of impact felt in Finland" (http://www2.hs.fi/english/archive/news.asp?id=20030411IE6). *Helsingin Sanomat*. 11 April 2003. . Retrieved 14 May 2008.
[97] Leyden, John (10 April 2003). "Nokia Networks axes 1,800 staff" (http://www.theregister.co.uk/2003/04/10/nokia_networks_axes/). *The Register*. . Retrieved 14 May 2008.
[98] "Nokia's Law (transcription)" (http://www.yle.fi/mot/kj050117/englishscript.htm). YLE TV1, Mot. 17 January 2005. . Retrieved 14 May 2008.
[99] "Nokia and Sanyo proposed new company will not proceed" (http://www.nokia.com/A4136002?newsid=1059331) (Press release). Nokia Corporation. 26 June 2006. . Retrieved 14 May 2008.
[100] "Nokia decides not to go forward with Sanyo CDMA partnership and plans broad restructuring of its CDMA business" (http://www.nokia.com/A4136002?newsid=1059329) (Press release). Nokia Corporation. 22 June 2006. . Retrieved 14 May 2008.

[101] "Nokia and Sanyo Announce Intent to Form a Global CDMA Mobile Phones Business" (http://www.nokia.com/A4136002?newsid=1034612) (Press release). Nokia Corporation. 14 February 2006. . Retrieved 14 May 2008.
[102] "Shell appoints Jorma Ollila as new Chairman" (http://www.shell.com/home/content/media/news_and_library/press_releases/2005/pr_announcement_04082005.html) (Press release). Royal Dutch Shell. 4 August 2005. . Retrieved 22 March 2009.
[103] "Nokia moves forward with management succession plan" (http://www.nokia.com/A4136002?newsid=1004430) (Press release). Nokia Corporation. 1 August 2005. . Retrieved 22 March 2009.
[104] Repo, Eljas; Melender, Tommi (19 September 2005). "Changing the guard at Nokia – Olli-Pekka Kallasvuo takes the helm" (http://finland.fi/netcomm/news/showarticle.asp?intNWSAID=41296&LAN=ENG). *Ministry for Foreign Affairs of Finland*. Virtual Finland. . Retrieved 22 March 2009.
[105] Kallasvuo, Olli-Pekka; President and CEO (8 May 2008). "2008 Nokia Annual General Meeting (transcription)" (http://nds1.nokia.com/NOKIA_COM_1/Microsites/AGM_2008/pdf/OPK_AGM_2008_ENGLISH.pdf) (PDF). Helsinki Fair Centre, Amfi Hall: Nokia Corporation. . Retrieved 12 June 2009.
[106] " □□□□□□□□ " (http://www.nokia.co.jp/about/release_081127.shtml) (in Japanese). – –. 27 November 2008. . Retrieved 5 December 2008.
[107] Nokia Will Lay off 4,000 and Move More Manufacturing to Asia | PCWorld Business Center (http://www.pcworld.com/businesscenter/article/249507/nokia_will_lay_off_4000_and_move_more_manufacturing_to_asia.html)
[108] Nokia Lays Off 1,000 Employees From Finnish Plant, Will Focus On Software (http://i2mag.com/nokia-lays-off-1000-employees-from-finnish-plant-will-focus-on-software/)
[109] "Nokia completes acquisition of assets of Sega.com Inc." (http://www.nokia.com/A4136002?newsid=918198) (Press release). Nokia Corporation. 22 September 2003. . Retrieved 16 March 2009.
[110] "Nokia to extend leadership in enterprise mobility with acquisition of Intellisync" (http://www.nokia.com/A4136002?newsid=1021663) (Press release). Nokia Corporation. 16 November 2005. . Retrieved 22 March 2009.
[111] "Nokia completes acquisition of Intellisync" (http://www.nokia.com/A4136002?newsid=1034184) (Press release). Nokia Corporation. 10 February 2006. . Retrieved 22 March 2009.
[112] "Nokia and Siemens to merge their communications service provider businesses" (http://www.nokia.com/A4136002?newsid=1057716) (Press release). Nokia Corporation. 19 June 2006. . Retrieved 22 March 2009.
[113] "Nokia to acquire Loudeye and launch a comprehensive mobile music experience" (http://www.nokia.com/A4136002?newsid=1067845) (Press release). Nokia Corporation. 8 August 2006. . Retrieved 14 May 2008.
[114] "Nokia completes Loudeye acquisition" (http://www.nokia.com/A4136001?newsid=1081455) (Press release). Nokia Corporation. 16 October 2006. . Retrieved 14 May 2008.
[115] "Nokia acquires Twango to offer a comprehensive media sharing experience" (http://www.nokia.com/A4136001?newsid=1141417) (Press release). Nokia Corporation. 24 July 2007. . Retrieved 14 May 2008.
[116] "Nokia Acquires Twango – Frequently Asked Questions (FAQ)" (http://www.nokia.com/NOKIA_COM_1/Press/Materials/NokiaTwangoFAQ.pdf) (PDF). Nokia Corporation. . Retrieved 14 May 2008.
[117] "Nokia to acquire Enpocket to create a global mobile advertising leader" (http://www.nokia.com/A4136001?newsid=1153772) (Press release). Nokia Corporation. 17 September 2007. . Retrieved 14 May 2008.
[118] Niccolai, James (1 October 2007). "Nokia buys mapping service for $8.1 billion" (http://www.infoworld.com/article/07/10/01/Nokia-buys-mapping-service-for-8.1-billion_1.html). *IDG News Service* (InfoWorld). . Retrieved 14 May 2008.
[119] "Nokia completes its acquisition of NAVTEQ" (http://www.nokia.com/A4136001?newsid=1235107) (Press release). Nokia Corporation. 10 July 2008. . Retrieved 22 March 2009.
[120] "Nokia to acquire leading consumer email and instant messaging provider OZ Communications" (http://news.taume.com/World-Business/Business-Finance/Nokia-to-acquire-leading-consumer-email-and-instant-messaging-provider-OZ-Communications-6922). *Taume News*. 30 September 2008. . Retrieved 30 September 2008.
[121] "Nokia to acquire cellity" (http://www.nokia.com/press/press-releases/showpressrelease?newsid=1330831) (Press release). Nokia Corporation. 24 July 2009. . Retrieved 4 August 2009.
[122] "Nokia completes acquisition of cellity" (http://www.nokia.com/press/press-releases/showpressrelease?newsid=1332884) (Press release). Nokia Corporation. 5 August 2009. . Retrieved 6 August 2009.
[123] "Nokia has acquired Plum" (http://www.nokia.com/press/press-releases/showpressrelease?newsid=1340931) (Press release). Nokia Corporation. 11 September 2009. . Retrieved 28 January 2010.
[124] "Nokia Acquires Browser Firm Novarra" (http://techie-buzz.com/mobile-news/nokia-acquires-browser-firm-novarra.html). Techie-buzz. 28 March 2010. . Retrieved 29 March 2010.
[125] "Nokia Acquires MetaCarta" (http://www.informationweek.com/news/mobility/smart_phones/showArticle.jhtml?articleID=224202519). informationweek. 11 April 2010. . Retrieved 12 April 2010.
[126] http://www.theverge.com/mobile/2012/1/7/2690366/nokia-buys-smarterphone-developer-of-feature-phone-operating-system
[127] http://www.pocket-lint.com/news/46116/nokia-acquires-scalado-for-better-image-quality
[128] "Nokia to cut 3500 jobs worldwide; to shut Romania factory" (http://www.moneycontrol.com/news/world-news/nokia-to-cut-3500-jobs-worldwide-to-shut-romania-factory_592235.html). 29 September 2011. .
[129] Moen, Arild (8 February 2012). "Nokia to Cut 4,000 Jobs" (http://online.wsj.com/article/SB10001424052970204136404577210401816583074.html). *The Wall Street Journal*. .

[130] "Nokia to cut 10,000 jobs; shut units" (http://www.thehindu.com/business/companies/article3528038.ece). *The Hindu*. 14 June 2012. .

[131] ben-Aaron, Diana (14 June 2012). "Nokia to Cut 10,000 Jobs as Elop Tries to Stanch Losses" (http://www.bloomberg.com/news/2012-06-14/nokia-to-cut-10-000-jobs-as-elop-tries-to-stanch-losses.html). *Bloomberg*. .

[132] http://timesofindia.indiatimes.com/tech/itslideshow/14151500.cms

[133] http://i.nokia.com/blob/view/-/263824/data/1/-/Request-Nokia-in-2010-pdf.pdf

[134] Nokia Downgraded to Junk - Zacks.com (http://www.zacks.com/stock/news/77208/nokia-downgraded-to-junk)

[135] "Nokia CEO Stephen Elop admits failure to foresee fast-changing industry" (http://economictimes.indiatimes.com/news/international-business/nokia-ceo-stephen-elop-admits-failure-to-foresee-fast-changing-industry/articleshow/14466105.cms). 28 June 2012. .

[136] "Structure" (http://www.nokia.com/about-nokia/company/structure). Nokia Corporation. 1 October 2009. . Retrieved 28 December 2009.

[137] "Nokia Siemens Networks starts operations and assumes a leading position in the communications industry" (http://www.nokia.com/A4136002?newsid=1116423) (Press release). Nokia Corporation. 2 April 2007. . Retrieved 7 April 2009.

[138] "Nokia's 25 percent profit jump falls short of expectations" (http://www.usatoday.com/money/economy/2008-04-17-173945271_x.htm). Associated Press. USA Today. 17 April 2008. . Retrieved 14 May 2008.

[139] "Worldwide Mobile Phone Market Maintains Its Growth Trajectory in the Fourth Quarter Despite Soft Demand for Feature Phones, According to IDC" (http://www.idc.com/getdoc.jsp?containerId=prUS23297412). ICD. 1 February 2012. .

[140] "The Wave of the Future" (http://www.underconsideration.com/brandnew/archives/the_wave_of_the_future.php). *Brand New: Opinions on Corporate and Brand Identity Work*. UnderConsideration LLC. 25 March 2007. . Retrieved 14 May 2008.

[141] "Reviews – 2007 – Nokia Siemens Networks" (http://www.identityworks.com/reviews/2007/Nokia_Siemens.htm). *Identityworks*. 2007. . Retrieved 14 May 2008.

[142] "Nokia Leadership Team" (http://www.nokia.com/A4126335). Nokia Corporation. April 2007. . Retrieved 14 May 2008.

[143] "Board of Directors" (http://www.nokia.com/A4126350). Nokia Corporation. April 2007. . Retrieved 14 May 2008.

[144] "Audit Committee Charter at Nokia" (http://www.nokia.com/NOKIA_COM_1/About_Nokia/Sidebars_new_concept/Board_charters/audit_charter.pdf) (PDF). Nokia Corporation. 2007. . Retrieved 14 May 2008.

[145] "Personnel Committee Charter at Nokia" (http://www.nokia.com/NOKIA_COM_1/About_Nokia/Sidebars_new_concept/Board_charters/personnel_charter_2007.pdf) (PDF). Nokia Corporation. 2007. . Retrieved 14 May 2008.

[146] "Corporate Governance and Nomination Committee Charter at Nokia" (http://www.nokia.com/NOKIA_COM_1/About_Nokia/Sidebars_new_concept/Board_charters/CG_Charter_2008_Final_20080123.pdf) (PDF). Nokia Corporation. 2008. . Retrieved 14 May 2008.

[147] "Committees of the Board" (http://www.nokia.com/link?cid=EDITORIAL_4207). Nokia Corporation. May 2007. . Retrieved 14 May 2008.

[148] Virkkunen, Johannes (29 September 2006). "New Finnish Companies Act designed to increase Finland's competitiveness" (http://www.lmr.fi/publications/companies_act_290906.pdf) (PDF). *LMR Attorneys Ltd. (Luostarinen Mettälä Räikkönen)*. . Retrieved 14 May 2008.

[149] "Articles of Association" (http://www.nokia.com/NOKIA_COM_1/About_Nokia/Company/Corporate_Governance/Articles_of_Association/Nokia_Articles_of_Association_10052007.pdf) (PDF). Nokia Corporation. 10 May 2007. . Retrieved 14 May 2008.

[150] "Corporate Governance Guidelines at Nokia" (http://www.nokia.com/NOKIA_COM_1/About_Nokia/Sidebars_new_concept/Board_charters/corporate_governance_guideline_sep06.pdf) (PDF). Nokia Corporation. 2006. . Retrieved 14 May 2008.

[151] "Suomalaisten yritysten ylin johto" (http://www.kolumbus.fi/taglarsson/dokumentit/yritys.htm) (in Finnish). . Retrieved 20 March 2009.

[152] "Nokia Research Center" (http://www.nokia.com/NOKIA_COM_1/Press/twwln/press_kit/Nokia_Research_Center_Press_Backgrounder_October_2007.pdf) (PDF). Nokia Corporation. October 2007. . Retrieved 14 May 2008.

[153] "About NRC – Nokia Research Center" (http://research.nokia.com/aboutus/index.html). Nokia Corporation. . Retrieved 17 March 2009.

[154] "NRC Locations – Nokia Research Center" (http://research.nokia.com/locations/index.html). Nokia Corporation. . Retrieved 17 March 2009.

[155] "INdT – Instituto Nokia de Tecnologia" (http://www.indt.org.br/). Nokia Corporation. . Retrieved 17 March 2009.

[156] "Production units" (http://www.nokia.com/A4149133). Nokia Corporation. June 2008. . Retrieved 14 May 2008.

[157] "Nokia despre sechestrul ANAF: Colaborăm pentru a ne asigura că situaţia e soluţionată satisfăcător" (http://www.mediafax.ro/economic/nokia-despre-sechestrul-anaf-colaboram-pentru-a-ne-asigura-ca-situatia-e-solutionata-satisfacator-8963720) (Press release). mediafax.ro. 12 November 2011. . Retrieved 12 November 2011.

[158] Kapanen, Ari (24 July 2007). "Ulkomaalaiset valtaavat pörssiyhtiöitä" (http://www.taloussanomat.fi/porssi-ja-raha/2007/07/24/Ulkomaalaiset+valtaavat+pörssiyhtiöitä/200717658/103) (in Finnish). *Taloussanomat*. . Retrieved 14 May 2008.

[159] "Nokia is no longer Finland's most valuable company" (http://www.phonearena.com/news/Nokia-is-no-longer-Finlands-most-valuable-company_id28750). phonearena.com. 4 April 2012. .

[160] Ali-Yrkkö, Jyrki (2001). "The role of Nokia in the Finnish Economy" (http://www.etla.fi/files/940_FES_01_1_nokia.pdf) (PDF). ETLA (The Research Institute of the Finnish Economy). . Retrieved 21 March 2009.

[161] Ali-Yrkkö, Jyrki (2010). "NOKIA AND FINLAND IN A SEA OF CHANGE" (http://www.etla.fi/files/2585_nokia_kirja_8_v2_kansineen.pdf). *ETLA – Research Institute of the Finnish Economy*. . Retrieved 12 November 2011.

[162] Guardian newspaper: Nokia cuts 4,000 jobs and moves smartphone manufacturing to Asia, 9 February 2012 (http://www.guardian.co.uk/technology/2012/feb/08/nokia-cuts-4000-manufacturing-jobs)

[163] "HS Archives" (http://www.hs.fi/arkisto/haku?pageNumber=1&order=FIFO&advancedSearch=&free=Connecting+People+Ove+Strandberg&date=year2003&depa=Kaikki+osastot&fromDay=0&fromMonth=0&fromYear=0&toDay=0&toMonth=0&toYear=0) (in Finnish). Helsingin Sanomat. 1 June 2003. . Retrieved 14 May 2008.
[164] "NOKIA | Connecting Pople 1992 Vector Logo (AI EPS)" (http://www.hdicon.com/vector-logos/nokia-connecting-pople-1992/). *HDicon.com*. . Retrieved 17 October 2010.
[165] Pitkänen, Juhani (Nokia's Art Director) (3 September 2007). "Nokia Strategic Marketing, Brand Identity" (http://www.nokia.com/A4126575). Nokia Corporation. . Retrieved 14 May 2008.
[166] "NOKIA | Connecting Pople new Vector Logo (AI EPS)" (http://www.hdicon.com/vector-logos/nokia-connecting-pople-new/). *HDicon.com*. . Retrieved 17 October 2010.
[167] "Erik Spiekermann – Furniture, Designs & Home Decor" (http://www.dwr.com/category/designers/r-t/erik-spiekermann.do). Design Within Reach. . Retrieved 7 January 2010.
[168] "Our New Typeface" (http://brandbook.nokia.com/blog/view/item62250/). Nokia Little Blog of Branding. . Retrieved 3 April 2012.
[169] http://thenextweb.com/eu/2012/04/04/poor-nokia-isnt-even-the-most-valuable-company-in-finland-anymore/
[170] http://media.corporate-ir.net/media_files/IROL/10/107224/Nokia_results2011Q2e.pdf
[171] http://www.theregister.co.uk/2012/04/19/nokia_earnings_ouch/
[172] http://thenextweb.com/insider/2012/05/14/nokia-is-getting-pummeled-stock-price-hits-staggering-low-after-a-6-nosedive/
[173] http://yle.fi/uutiset/nokia_share_price_in_nose_dive_to_2_euros/6094843
[174] "Nokia Way and values" (http://www.nokia.com/A4126303). Nokia Corporation. . Retrieved 14 May 2008.
[175] "dotMobi Investors" (http://web.archive.org/web/20080507214157/http://mtld.mobi/company/about/investors). dotMobi. Archived from the original (http://mtld.mobi/company/about/investors) on 7 May 2008. . Retrieved 14 May 2008.
[176] Haumont, Serge; Siren, Ritva. "dotMobi, a Key Enabler for the Mobile Internet" (http://research.nokia.com/files/Haumont-dotMobi.pdf) (PDF). *Nokia Research Center*. Nokia Corporation. . Retrieved 14 May 2008.
[177] "Nokia Ad Business" (http://www.adservice.nokia.com/faq.jsp#12). Nokia Corporation. . Retrieved 14 May 2008.
[178] Reardon, Marguerite (6 March 2007). "Nokia introduces mobile ad services" (http://news.com.com/2100-1039_3-6164800.html). *CNET News.com*. . Retrieved 14 May 2008.
[179] "Meet Ovi, the door to Nokia's Internet services" (http://www.nokia.com/A4136001?newsid=1149749) (Press release). Nokia Corporation. 29 August 2007. . Retrieved 7 April 2009.
[180] Niccolai, James (4 December 2007). "Nokia Lays Plan for More Internet Services" (http://www.nytimes.com/idg/IDG_002570DE00740E18002573A70046F2EF.html?ref=technology). *IDG News Service* (New York Times). . Retrieved 14 May 2008.
[181] "Ovi by Nokia" (http://www.nokia.com/NOKIA_COM_1/Press/Materials/White_Papers/pdf_files/backgrounders2008/Backgrounder_Ovi_by_Nokia.pdf) (PDF). Nokia Corporation. . Retrieved 7 April 2009.
[182] "Ovi Store opens for business" (http://www.nokia.com/press/press-releases/showpressrelease?newsid=1317441) (Press release). Nokia Corporation. 26 May 2009. . Retrieved 12 June 2009.
[183] Virki, Tarmo (18 March 2009). "Nokia to shutter its "Mosh" success story" (http://www.reuters.com/article/technologyNews/idUSTRE52H6AI20090318?pageNumber=1&virtualBrandChannel=0). *Reuters*. . Retrieved 14 July 2009.
[184] "Nokia launches new online magazine" (http://www.newstatesman.com/magazines/2010/03/online-magazine-nokia-ovi). NewStatesman. 23 March 2010. . Retrieved 29 March 2010.
[185] "Nokia – My Nokia" (http://europe.nokia.com/my-nokia). Nokia Corporation. . Retrieved 10 January 2012.
[186] "Nokia Retreats from Music Service" (http://www.yle.fi/uutiset/news/2011/01/nokia_retreats_from_music_service_2294706.html). YLE. 18 January 2011. . Retrieved 18 January 2011.
[187] Fields, Davis (17 December 2008). "Nokia Email service graduates as part of Nokia Messaging" (http://betalabs.nokia.com/blog/2008/12/17/nokia-email-service-graduates-as-part-of-nokia-messaging/). *Nokia Beta Labs*. Nokia Corporation. . Retrieved 16 March 2009.
[188] "Nokia Messaging: FAQ" (http://email.nokia.com/account/faq.action?change_locale=en). Nokia Corporation. . Retrieved 12 June 2009.
[189] Cellan-Jones, Rory (22 June 2009). "Hi-tech helps Iranian monitoring" (http://news.bbc.co.uk/1/hi/technology/8112550.stm). *BBC News*. . Retrieved 14 July 2009.
[190] Rhoads, Christopher; Chao, Loretta (22 June 2009). "Iran's Web Spying Aided By Western Technology" (http://online.wsj.com/article/SB124562668777335653.html#mod). *The Wall Street Journal* (Dow Jones & Company, Inc.): pp. A1. . Retrieved 14 July 2009.
[191] "Provision of Lawful Intercept capability in Iran" (http://www.nokiasiemensnetworks.com/global/Press/Press+releases/news-archive/Provision+of+Lawful+Intercept+capability+in+Iran.htm) (Press release). Nokia Siemens Networks. 22 June 2009. . Retrieved 14 July 2009.
[192] Kamali Dehghan, Saeed (14 July 2009). "Iranian consumers boycott Nokia for 'collaboration'" (http://www.guardian.co.uk/world/2009/jul/14/nokia-boycott-iran-election-protests). *The Guardian* (London: Guardian News and Media Limited). . Retrieved 27 July 2009.
[193] Ozimek, John (6 March 2009). "'Lex Nokia' company snoop law passes in Finland" (http://www.theregister.co.uk/2009/03/06/finland_nokia_snooping/). *The Register*. . Retrieved 27 July 2009.
[194] "Nokia Denies Threat to Leave Finland" (http://www.cellular-news.com/story/35783.php). *cellular-news*. 1 February 2009. . Retrieved 27 July 2009.
[195] Virki, Tarmo (18 January 2010). "SCENARIOS-What lies ahead in Nokia vs Apple legal battle" (http://www.reuters.com/article/idUSLDE60H05R20100118?type=marketsNews). *Reuters*. . Retrieved 25 January 2010.

[196] "The war of the Smartphones: Nokia's new patent suit against Apple" (http://pda-phone-reviews.in/latest-news/nokias-new-patent-suit-against-apple/). *Snartphone Reviews*. 6 January 2010. . Retrieved 25 January 2010.
[197] "Nokia's Patent Settlement With Apple Won't Help Much" (http://www.informationweek.com/news/personal-tech/smart-phones/230600172). 14 June 2011. . Retrieved 29 June 2011.
[198] Smith, Catharine (14 June 2011). "Apple Settles With Nokia In Patent Lawsuit" (http://www.huffingtonpost.com/2011/06/14/apple-nokia-patent-lawsuit-settlement_n_876499.html). *Huffington Post*. . Retrieved 29 June 2011.
[199] ben-Aaron, Diana; Pohjanpalo, Kati (14 June 2011). "Nokia Wins Apple Patent-License Deal Cash, Settles Lawsuits" (http://www.bloomberg.com/news/2011-06-14/nokia-apple-payments-to-nokia-settle-all-litigation.html). *Bloomberg*. . Retrieved 29 June 2011.
[200] "Guide to Greener Electronics – Greenpeace International" (http://www.greenpeace.org/international/en/campaigns/climate-change/cool-it/Guide-to-Greener-Electronics/). Greenpeace International. . Retrieved 14 November 2011.
[201] "Nokia – Where and how to recycle – Recycling – Environment" (http://www.nokia.com/environment/recycling/where-and-how-to-recycle). Nokia. . Retrieved 12 August 2010.
[202] "Global consumer survey reveals that majority of old mobile phones are lying in drawers at home and not being recycled" (http://www.nokia.com/press/press-releases/showpressrelease?newsid=1234291) (Press release). Nokia Corporation. 8 July 2008. . Retrieved 27 July 2009.
[203] "Nokia – Energy efficiency – Devices and services – Environment" (http://www.nokia.com/environment/we-energise/nokia-and-energy-efficiency). Nokia. . Retrieved 12 August 2010.
[204] "Nokia – Energy saving targets – Environmental strategy – Strategy and reports – Environment" (http://www.nokia.com/environment/strategy-and-reports/environmental-strategy/energy-saving-targets). Nokia. . Retrieved 12 August 2010.
[205] "Nokia" (http://www.greenpeace.org/international/en/campaigns/toxics/electronics/Guide-to-Greener-Electronics/companies/Nokia/). Greenpeace International. . Retrieved 12 August 2010.
[206] "Materials and substances" (http://www.nokia.com/environment/we-create/materials-and-substances). Nokia Corporation. . Retrieved 12 August 2010.
[207] "Eco declarations" (http://www.nokia.com/environment/we-create/devices-and-accessories/eco-declarations). Nokia Corporation. . Retrieved 27 July 2009.
[208] Rubio, Jenalyn (12 April 2008). "Tech Goes Greener" (http://www.pcworld.com/article/id,144482-c,recycling/article.html). *Computerworld Philippines* (PC World). . Retrieved 14 May 2008.
[209] "Nokia Remade Concept Phone goes Green" (http://www.mobiletor.com/2008/04/09/nokia-remade-concept-phone-goes-green/). Mobiletor. 9 April 2008. . Retrieved 14 May 2008.
[210] "Open Innovation – Nokia Research Center" (http://research.nokia.com/openinnovation). Nokia Corporation. . Retrieved 1 April 2009.
[211] "Nokia is India's most trusted brand: Brand Trust" (http://profit.ndtv.com/news/show/nokia-is-india-s-most-trusted-brand-brand-trust-136783). profit.ndtv.com. 19 January 2011. . Retrieved 21 November 2011.

Further reading

Title	Author	Publisher	Year	Length	ISBN
Winning Across Global Markets: How Nokia Creates Strategic Advantage in a Fast-Changing World	Dan Steinbock	Jossey-Bass / Wiley	May 2010	304 pp	ISBN 978-0-470-33966-4
Nokia: The Inside Story	Martti Häikiö	FT / Prentice Hall	October 2002	256 pp	ISBN 0-273-65983-9
Work Goes Mobile: Nokia's Lessons from the Leading Edge	Michael Lattanzi, Antti Korhonen, Vishy Gopalakrishnan	John Wiley & Sons	January 2006	212 pp	ISBN 0-470-02752-5
Mobile Usability: How Nokia Changed the Face of the Mobile Phone	Christian Lindholm, Turkka Keinonen, Harri Kiljander	McGraw-Hill Companies	June 2003	301 pp	ISBN 0-07-138514-2
Business The Nokia Way: Secrets of the World's Fastest Moving Company	Trevor Merriden	John Wiley & Sons	February 2001	168 pp	ISBN 1-84112-104-5
The Nokia Revolution: The Story of an Extraordinary Company That Transformed an Industry	Dan Steinbock	AMACOM Books	April 2001	375 pp	ISBN 0-8144-0636-X

External links

- Official Nokia international website (http://www.nokia.com/)

Wi-Fi

Wi-Fi (/ˈwaɪfaɪ/, also spelled *Wifi* or *WiFi*) is a popular technology that allows an electronic device to exchange data wirelessly (using radio waves) over a computer network, including high-speed Internet connections. The Wi-Fi Alliance defines Wi-Fi as any "wireless local area network (WLAN) products that are based on the Institute of Electrical and Electronics Engineers' (IEEE) 802.11 standards".[1] However, since most modern WLANs are based on these standards, the term "Wi-Fi" is used in general English as a synonym for "WLAN".

Wi-Fi logo

A device that can use Wi-Fi (such as a personal computer, video game console, smartphone, tablet, or digital audio player) can connect to a network resource such as the Internet via a wireless network access point. Such an access point (or hotspot) has a range of about 20 meters (65 feet) indoors and a greater range outdoors. Hotspot coverage can comprise an area as small as a single room with walls that block radio waves or as large as many square miles — this is achieved by using multiple overlapping access points.

"Wi-Fi" is a trademark of the Wi-Fi Alliance and the brand name for products using the IEEE 802.11 family of standards. Only Wi-Fi products that complete Wi-Fi Alliance interoperability certification testing successfully may use the "Wi-Fi CERTIFIED" designation and trademark.

Wi-Fi has had a checkered security history. Its earliest encryption system, WEP, proved easy to break. Much higher quality protocols, WPA and WPA2, were added later. However, an optional feature added in 2007, called Wi-Fi Protected Setup (WPS), has a flaw that allows a remote attacker to recover the router's WPA or WPA2 password in a few hours on most implementations.[2] Some manufacturers have recommended turning off the WPS feature. The Wi-Fi Alliance has since updated its test plan and certification program to ensure all newly-certified devices resist brute-force AP PIN attacks.

History

802.11 technology has its origins in a 1985 ruling by the US Federal Communications Commission that released the ISM band for unlicensed use.[3] In 1991, NCR Corporation with AT&T invented the precursor to 802.11 intended for use in cashier systems. The first wireless products were under the name WaveLAN.

Vic Hayes has been called the "father of Wi-Fi". He was involved in designing the initial standards within the IEEE.[4][5]

A large number of patents by many companies are used in 802.11 standard.[6] In 1992 and 1996, Australian organisation CSIRO obtained patents for a method later used in Wi-Fi to "unsmear" the signal.[7] In April 2009, 14 tech companies agreed to pay CSIRO $250 million for infringements on CSIRO patents.[8] This led to WiFi being attributed as an Australian invention,[9] though this has been the subject of some controversy.[10][11] CSIRO won a further $220 million settlement for Wi-Fi patent infringements in 2012 with global firms in the United States required to pay the CSIRO licensing rights estimated to be worth an additional $1 billion in royalties.[8][12][13]

In 1999, the Wi-Fi Alliance was formed as a trade association to hold the Wi-Fi trademark under which most products are sold.[14]

The name

The term *Wi-Fi*, first used commercially in August 1999,[15] was coined by a brand-consulting firm called Interbrand Corporation. The Wi-Fi Alliance had hired Interbrand to determine a name that was "a little catchier than 'IEEE 802.11b Direct Sequence'".[16][17][18] Belanger also stated that Interbrand invented *Wi-Fi* as a play on words with *Hi-Fi* (high fidelity), and also created the Wi-Fi logo.

The Wi-Fi Alliance initially used an advertising slogan for Wi-Fi, "The Standard for Wireless Fidelity",[16] but later removed the phrase from their marketing. Despite this, some documents from the Alliance dated 2003 and 2004 still contain the term *Wireless Fidelity*.[19][20] There was no official statement related to the dropping of the term.

The yin-yang Wi-Fi logo indicates the certification of a product for interoperability.[19]

Non-Wi-Fi technologies intended for fixed points such as Motorola Canopy are usually described as fixed wireless. Alternative wireless technologies include mobile phone standards such as 2G, 3G or 4G.

Wi-Fi certification

The IEEE does not test equipment for compliance with their standards. The non-profit Wi-Fi Alliance was formed in 1999 to fill this void — to establish and enforce standards for interoperability and backward compatibility, and to promote wireless local-area-network technology. As of 2010, the Wi-Fi Alliance consisted of more than 375 companies from around the world.[21][22] The Wi-Fi Alliance enforces the use of the Wi-Fi brand to technologies based on the IEEE 802.11 standards from the Institute of Electrical and Electronics Engineers. This includes wireless local area network (WLAN) connections, device to device connectivity (such as Wi-Fi Peer to Peer aka Wi-Fi Direct), Personal area network (PAN), local area network (LAN) and even some limited wide area network (WAN) connections. Manufacturers with membership in the Wi-Fi Alliance, whose products pass the certification process, gain the right to mark those products with the Wi-Fi logo.

Specifically, the certification process requires conformance to the IEEE 802.11 radio standards, the WPA and WPA2 security standards, and the EAP authentication standard. Certification may optionally include tests of IEEE 802.11 draft standards, interaction with cellular-phone technology in converged devices, and features relating to security set-up, multimedia, and power-saving.[23]

Not every Wi-Fi device is submitted for certification. The lack of Wi-Fi certification does not necessarily imply a device is incompatible with other Wi-Fi devices. If it is compliant or partly compatible, the Wi-Fi Alliance may not object to its description as a Wi-Fi device though technically only certified devices are approved. Derivative terms, such as Super Wi-Fi, coined by the US Federal Communications Commission (FCC) to describe proposed networking in the UHF TV band in the US, may or may not be sanctioned.

Uses

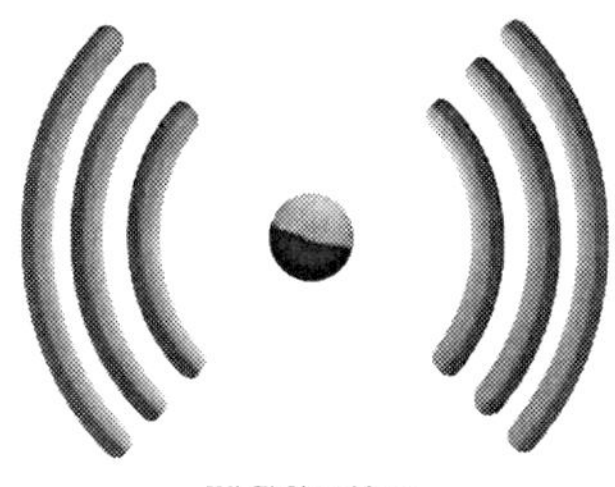

Wi-Fi Signal logo

To connect to a Wi-Fi LAN, a computer has to be equipped with a wireless network interface controller. The combination of computer and interface controller is called a *station*. All stations share a single radio frequency communication channel. Transmissions on this channel are received by all stations within range. The hardware does not signal the user that the transmission was delivered and is therefore called a best-effort delivery mechanism. A carrier wave is used to transmit the data in packets, referred to as "Ethernet frames". Each station is constantly tuned in on the radio frequency communication channel to pick up available transmissions.

Internet access

A Wi-Fi-enabled device can connect to the Internet when within range of a wireless network connected to the Internet. The coverage of one or more (interconnected) access points — called hotspots — can extend from an area as small as a few rooms to as large as many square miles. Coverage in the larger area may require a group of access points with overlapping coverage. Outdoor public Wi-Fi technology has been used successfully in wireless mesh networks in London, UK.

Wi-Fi provides service in private homes, high street chains and independent businesses, as well as in public spaces at Wi-Fi hotspots set up either free-of-charge or commercially. Organizations and businesses, such as airports, hotels, and restaurants, often provide free-use hotspots to attract customers. Enthusiasts or authorities who wish to provide services or even to promote business in selected areas sometimes provide free Wi-Fi access.

Routers that incorporate a digital subscriber line modem or a cable modem and a Wi-Fi access point, often set up in homes and other buildings, provide Internet access and internetworking to all devices connected to them, wirelessly or via cable. With the emergence of MiFi and WiBro (a portable Wi-Fi router) people can easily create their own Wi-Fi hotspots that connect to Internet via cellular networks. Now Android, Bada, iOS (iPhone), and Symbian devices can create wireless connections.[24] Wi-Fi also connects places that normally don't have network access, such as kitchens and garden sheds.

City-wide Wi-Fi

Further information: Municipal wireless network

In the early 2000s, many cities around the world announced plans to construct city-wide Wi-Fi networks. There are many successful examples; in 2004, Mysore became India's first Wi-fi-enabled city and second in the world after Jerusalem. A company called WiFiyNet has set up hotspots in Mysore, covering the complete city and a few nearby villages.[25]

An outdoor Wi-Fi access point

In 2005, Sunnyvale, California, became the first city in the United States to offer city-wide free Wi-Fi,[26] and Minneapolis has generated $1.2 million in profit annually for its provider.[27]

In May 2010, London, UK, Mayor Boris Johnson pledged to have London-wide Wi-Fi by 2012.[28] Several boroughs including Westminster and Islington[29][30] already have extensive outdoor Wi-Fi coverage.

Officials in South Korea's capital are moving to provide free Internet access at more than 10,000 locations around the city, including outdoor public spaces, major streets and densely populated residential areas. Seoul will grant leases to KT, LG Telecom and SK Telecom. The companies will invest $44 million in the project, which will be completed in 2015.[31]

Campus-wide Wi-Fi

Many traditional college campuses in the United States provide at least partial wireless Wi-Fi Internet coverage. Carnegie Mellon University built the first campus-wide wireless Internet network, called Wireless Andrew at its Pittsburgh campus in 1993 before Wi-Fi branding originated.[32][33][34]

In 2000, Drexel University in Philadelphia became the United States's first major university to offer completely wireless Internet access across its entire campus.[35]

Direct computer-to-computer communications

Wi-Fi also allows communications directly from one computer to another without an access point intermediary. This is called *ad hoc* Wi-Fi transmission. This wireless ad hoc network mode has proven popular with multiplayer handheld game consoles, such as the Nintendo DS, Playstation Portable, digital cameras, and other consumer electronics devices. Some devices can also share their Internet connection using ad-hoc, becoming hotspots or "virtual routers".[36]

Similarly, the Wi-Fi Alliance promotes a specification called *Wi-Fi Direct* for file transfers and media sharing through a new discovery- and security-methodology.[37] Wi-Fi Direct launched in October 2010.[38]

Advantages and limitations

Advantages

A keychain-size Wi-Fi detector

Wi-Fi allows cheaper deployment of local area networks (LANs). Also spaces where cables cannot be run, such as outdoor areas and historical buildings, can host wireless LANs.

Manufacturers are building wireless network adapters into most laptops. The price of chipsets for Wi-Fi continues to drop, making it an economical networking option included in even more devices.

Different competitive brands of access points and client network-interfaces can inter-operate at a basic level of service. Products designated as "Wi-Fi Certified" by the Wi-Fi Alliance are backwards compatible. Unlike mobile phones, any standard Wi-Fi device will work anywhere in the world.

Wi-Fi Protected Access encryption (WPA2) is considered secure, provided a strong passphrase is used. New protocols for quality-of-service (WMM) make Wi-Fi more suitable for latency-sensitive applications (such as voice and video). Power saving mechanisms (WMM Power Save) extend battery life.

Limitations

Spectrum assignments and operational limitations are not consistent worldwide: most of Europe allows for an additional two channels beyond those permitted in the US for the 2.4 GHz band (1–13 vs. 1–11), while Japan has one more on top of that (1–14). As of 2007, Europe, is essentially homogeneous in this respect.

A Wi-Fi signal occupies five channels in the 2.4 GHz band. Any two channels numbers that differ by five or more, such as 2 and 7, do not overlap. The oft-repeated adage that channels 1, 6, and 11 are the *only* non-overlapping channels is, therefore, not accurate. Channels 1, 6, and 11 are the only *group of three* non-overlapping channels in the U.S.

Equivalent isotropically radiated power (EIRP) in the EU is limited to 20 dBm (100 mW).

The current 'fastest' norm, 802.11n, uses double the radio spectrum/bandwidth (40 MHz) compared to 802.11a or 802.11g (20 MHz). This means there can be only one 802.11n network on the 2.4 GHz band at a given location, without interference to/from other WLAN traffic. 802.11n can also be set to use 20 MHz bandwidth only to prevent interference in dense community.

Range

Wi-Fi networks have limited range. A typical wireless access point using 802.11b or 802.11g with a stock antenna might have a range of 32 m (120 ft) indoors and 95 m (300 ft) outdoors. IEEE 802.11n, however, can more than double the range.[39] Range also varies with frequency band. Wi-Fi in the 2.4 GHz frequency block has slightly better range than Wi-Fi in the 5 GHz frequency block which is used by 802.11a and optionally by 802.11n. On wireless routers with detachable antennas, it is possible to improve range by fitting upgraded antennas which have higher gain in particular directions. Outdoor ranges can be improved to many kilometers through the use of high gain directional antennas at the router and remote device(s). In general, the maximum amount of power that a Wi-Fi device can transmit is limited by local regulations, such as FCC Part 15 in the US.

Due to reach requirements for wireless LAN applications, Wi-Fi has fairly high power consumption compared to some other standards. Technologies such as Bluetooth (designed to support wireless PAN applications) provide a much shorter propagation range of <10m[40] and so in general have a lower power consumption. Other low-power technologies such as ZigBee have fairly long range, but much lower data rate. The high power consumption of Wi-Fi makes battery life in mobile devices a concern.

Researchers have developed a number of "no new wires" technologies to provide alternatives to Wi-Fi for applications in which Wi-Fi's indoor range is not adequate and where installing new wires (such as CAT-5) is not possible or cost-effective. For example, the ITU-T G.hn standard for high speed Local area networks uses existing home wiring (coaxial cables, phone lines and power lines). Although G.hn does not provide some of the advantages of Wi-Fi (such as mobility or outdoor use), it's designed for applications (such as IPTV distribution) where indoor range is more important than mobility.

Due to the complex nature of radio propagation at typical Wi-Fi frequencies, particularly the effects of signal reflection off trees and buildings, algorithms can only approximately predict Wi-Fi signal strength for any given area in relation to a transmitter.[41] This effect does not apply equally to long-range Wi-Fi, since longer links typically operate from towers that transmit above the surrounding foliage.

The practical range of Wi-Fi essentially confines mobile use to such applications as inventory-taking machines in warehouses or in retail spaces, barcode-reading devices at check-out stands, or receiving/shipping stations. Mobile use of Wi-Fi over wider ranges is limited, for instance, to uses such as in an automobile moving from one hotspot to another. Other wireless technologies are more suitable for communicating with moving vehicles.

Data security risks

The most common wireless encryption-standard, Wired Equivalent Privacy (WEP), has been shown to be easily breakable even when correctly configured. Wi-Fi Protected Access (WPA and WPA2) encryption, which became available in devices in 2003, aimed to solve this problem. Wi-Fi access points typically default to an encryption-free (*open*) mode. Novice users benefit from a zero-configuration device that works out-of-the-box, but this default does not enable any wireless security, providing open wireless access to a LAN. To turn security on requires the user to configure the device, usually via a software graphical user interface (GUI). On unencrypted Wi-Fi networks connecting devices can monitor and record data (including personal information). Such networks can only be secured by using other means of protection, such as a VPN or secure Hypertext Transfer Protocol (HTTPS) over Transport Layer Security.

Interference

Wi-Fi connections can be disrupted or the internet speed lowered by having other devices in the same area. Many 2.4 GHz 802.11b and 802.11g access-points default to the same channel on initial startup, contributing to congestion on certain channels. Wi-Fi pollution, or an excessive number of access points in the area, especially on the neighboring channel, can prevent access and interfere with other devices' use of other access points, caused by overlapping channels in the 802.11g/b spectrum, as well as with decreased signal-to-noise ratio (SNR) between access points. This can become a problem in high-density areas, such as large apartment complexes or office buildings with many Wi-Fi access points.

Additionally, other devices use the 2.4 GHz band: microwave ovens, ISM band devices, security cameras, ZigBee devices, Bluetooth devices, video senders, cordless phones, baby monitors, and (in some countries) Amateur radio all of which can cause significant additional interference. It is also an issue when municipalities[42] or other large entities (such as universities) seek to provide large area coverage.

Hardware

Standard devices

A wireless access point (WAP) connects a group of wireless devices to an adjacent wired LAN. An access point resembles a network hub, relaying data between connected wireless devices in addition to a (usually) single connected wired device, most often an ethernet hub or switch, allowing wireless devices to communicate with other wired devices.

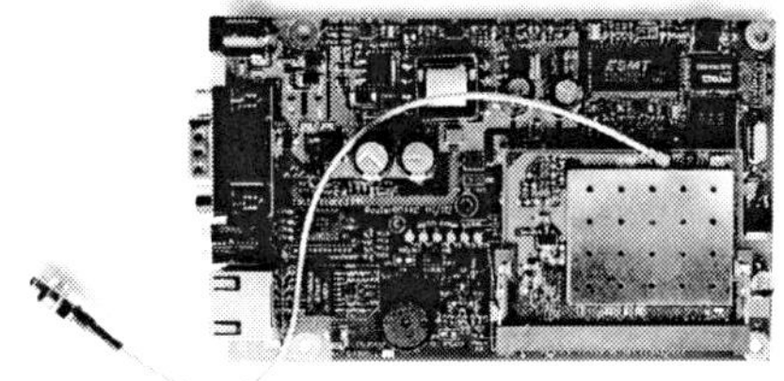

An embedded RouterBoard 112 with U.FL-RSMA pigtail and R52 mini PCI Wi-Fi card widely used by wireless Internet service providers (WISPs) in the Czech Republic

Wireless adapters allow devices to connect to a wireless network. These adapters connect to devices using various external or internal interconnects such as PCI, miniPCI, USB, ExpressCard, Cardbus and PC Card. As of 2010, most newer laptop computers come equipped with built in internal adapters.

OSBRiDGE 3GN - 802.11n Access Point and UMTS/GSM Gateway in one device

Wireless routers integrate a Wireless Access Point, ethernet switch, and internal router firmware application that provides IP routing, NAT, and DNS forwarding through an integrated WAN-interface. A wireless router allows wired and wireless ethernet LAN devices to connect to a (usually) single WAN device such as a cable modem or a DSL modem. A wireless router allows all three devices, mainly the access point and router, to be configured through one central utility. This utility is usually an integrated web server that is accessible to wired and wireless LAN clients and often optionally to WAN clients. This utility may also be an application that is run on a desktop computer, as is the case with as Apple's AirPort, which is managed with the AirPort Utility on Mac OS X and Microsoft Windows.[43]

Wireless network bridges connect a wired network to a wireless network. A bridge differs from an access point: an access point connects wireless devices to a wired network at the data-link layer. Two wireless bridges may be used to connect two wired networks over a wireless link, useful in situations where a wired connection may be unavailable, such as between two separate homes.

An Atheros Wi-Fi N draft adaptor with built in Bluetooth on a Sony Vaio E series laptop

Wireless range-extenders or wireless repeaters can extend the range of an existing wireless network. Strategically placed range-extenders can elongate a signal area or allow for the signal area to reach around barriers such as those pertaining in L-shaped corridors. Wireless devices connected through repeaters will suffer from an increased latency for each hop. Additionally, a wireless device connected to any of the repeaters in the chain will have a throughput limited by the "weakest link" between the two nodes in the chain from which the connection originates to where the connection ends.

USB wireless adapter

The security standard, Wi-Fi Protected Setup, allows embedded devices with limited graphical user interface to connect to the Internet with ease. Wi-Fi Protected Setup has 2 configurations: The Push Button configuration and the PIN configuration. These embedded devices are also called The Internet of Things and are low-power, battery-operated embedded systems. A number of Wi-Fi manufacturers design chips and modules for embedded Wi-Fi, such as GainSpan.[44]

Distance records

Distance records (using non-standard devices) include 382 km (237 mi) in June 2007, held by Ermanno Pietrosemoli and EsLaRed of Venezuela, transferring about 3 MB of data between the mountain-tops of El Águila and Platillon.[45][46] The Swedish Space Agency transferred data 420 km (260 mi), using 6 watt amplifiers to reach an overhead stratospheric balloon.[47]

Embedded systems

Increasingly in the last few years (particularly as of 2007), embedded Wi-Fi modules have become available that incorporate a real-time operating system and provide a simple means of wirelessly enabling any device which has and communicates via a serial port.[48] This allows the design of simple monitoring devices. An example is a portable ECG device monitoring a patient at home. This Wi-Fi-enabled device can communicate via the Internet.[49]

Embedded serial-to-Wi-Fi module

These Wi-Fi modules are designed by OEMs so that implementers need only minimal Wi-Fi knowledge to provide Wi-Fi connectivity for their products.

Multiple access points

Increasing the number of Wi-Fi access points provides network redundancy, support for fast roaming and increased overall network-capacity by using more channels or by defining smaller cells. Wi-Fi implementations have moved toward "thin" access points, with more of the network intelligence housed in a centralized network appliance, relegating individual access points to the role of "dumb" transceivers. Outdoor applications may use mesh topologies.

Network security

The main issue with wireless network security is its simplified access to the network compared to traditional wired networks such as ethernet. With wired networking one must either gain access to a building (physically connecting into the internal network) or break through an external firewall. Most business networks protect sensitive data and systems by attempting to disallow external access. Enabling wireless connectivity reduces security if the network uses inadequate or no encryption.[50]

An attacker who has gained access to a Wi-Fi network router can initiate a DNS spoofing attack against any other user of the network by forging a response before the queried DNS server has a chance to reply.[51]

Securing methods

A common measure to deter unauthorized users involves hiding the access point's name by disabling the SSID broadcast. While effective against the casual user, it is ineffective as a security method because the SSID is broadcast in the clear in response to a client SSID query. Another method is to only allow computers with known MAC addresses to join the network,[52] but determined eavesdroppers may be able join the network by spoofing an authorized address.

Wired Equivalent Privacy (WEP) encryption was designed to protect against casual snooping but it is no longer considered secure. Tools such as AirSnort or Aircrack-ng can quickly recover WEP encryption keys.[53] Because of WEP's weakness the Wi-Fi Alliance approved Wi-Fi Protected Access (WPA) which uses TKIP. WPA was specifically designed to work with older equipment usually through a firmware upgrade. Though more secure than WEP, WPA has known vulnerabilities.

The more secure WPA2 using Advanced Encryption Standard was introduced in 2004 and is supported by most new Wi-Fi devices. WPA2 is fully compatible with WPA.[54]

A flaw in a feature added to Wi-Fi in 2007, called Wi-Fi Protected Setup, allows WPA and WPA2 security to be bypassed and effectively broken in many situations. The only remedy as of late 2011 is to turn off Wi-Fi Protected Setup,[55] which is not always possible.

Piggybacking

Piggybacking refers to access to a wireless Internet connection by bringing one's own computer within the range of another's wireless connection, and using that service without the subscriber's explicit permission or knowledge.

During the early popular adoption of 802.11, providing open access points for anyone within range to use was encouraged to cultivate wireless community networks,[56] particularly since people on average use only a fraction of their downstream bandwidth at any given time.

Recreational logging and mapping of other people's access points has become known as wardriving. Indeed, many access points are intentionally installed without security turned on so that they can be used as a free service. Providing access to one's Internet connection in this fashion may breach the Terms of Service or contract with the ISP. These activities do not result in sanctions in most jurisdictions; however, legislation and case law differ considerably across the world. A proposal to leave graffiti describing available services was called warchalking.[57] A Florida court case determined that owner laziness was not to be a valid excuse.

Piggybacking often occurs unintentionally, since most access points are configured without encryption by default and operating systems can be configured to connect automatically to any available wireless network. A user who happens to start up a laptop in the vicinity of an access point may find the computer has joined the network without any visible indication. Moreover, a user intending to join one network may instead end up on another one if the latter has a stronger signal. In combination with automatic discovery of other network resources (see DHCP and Zeroconf) this could possibly lead wireless users to send sensitive data to the wrong middle-man when seeking a destination (*see Man-in-the-middle attack*). For example, a user could inadvertently use an unsecure network to log in to a website, thereby making the login credentials available to anyone listening, if the website uses an unsecure protocol such as HTTP.

Safety

Further information: Wireless electronic devices and health

The World Health Organization (WHO) says "there is no risk from low level, long-term exposure to wi-fi networks" and the United Kingdom's Health Protection Agency reports that exposure to Wi-Fi for a year results in "same amount of radiation from a 20-minute mobile phone call." [58][59]

A small percentage of Wi-Fi users have reported adverse health issues after repeat exposure and use of Wi-Fi,[60] though there has been no publication of any effects being observable in double-blind studies. A review of studies involving 725 people that claimed electromagnetic hypersensitivity found no evidence for their claims.[61]

One study speculated that "laptops (Wi-Fi mode) on the lap near the testicles may result in decreased male fertility".[62] Another study found decreased working memory among males during Wi-Fi exposure.[63]

Notes

[1] What is Wi-Fi? - A Word Definition From the Webopedia Computer Dictionary (http://www.webopedia.com/TERM/W/Wi_Fi.html)

[2] http://sviehb.files.wordpress.com/2011/12/viehboeck_wps.pdf

[3] "Wi-Fi (wireless networking technology)" (http://www.britannica.com/EBchecked/topic/1473553/Wi-Fi). *Encyclopædia Britannica*. . Retrieved 2010-02-03.

[4] Ben Charny (December 6, 2002). "CNET Vision series" (http://news.cnet.com/1200-1070-975460.html). CNET. . Retrieved 2011-10-14.

[5] Olga Kharif (April 1, 2003). "Paving the Airwaves for Wi-Fi" (http://www.businessweek.com/technology/content/apr2003/tc2003041_5423_tc107.htm). Bloomberg Businessweek. . Retrieved 2011-10-14.

[6] IEEE-SA - IEEE 802.11 and Amendments Patent Letters of Assurance (http://standards.ieee.org/about/sasb/patcom/pat802_11.html)

[7] David Sygall. " How Australia's top scientist earned millions from Wi-Fi (http://www.smh.com.au/technology/sci-tech/how-australias-top-scientist-earned-millions-from-wifi-20091207-kep4.html)". *The Sydney Morning Herald*, December 7, 2009.

[8] Moses, Asher (June 1, 2010). "CSIRO to reap 'lazy billion' from world's biggest tech companies" (http://www.theage.com.au/technology/enterprise/csiro-to-reap-lazy-billion-from-worlds-biggest-tech-companies-20100601-wsu2.html). *The Age* (Melbourne). . Retrieved 8 June 2010.

[9] World changing Aussie inventions - Australian Geographic (http://www.australiangeographic.com.au/journal/world-changing-aussie-inventions.htm)

[10] How the Aussie government "invented WiFi" and sued its way to $430 million | Ars Technica (http://arstechnica.com/tech-policy/news/2012/04/how-the-aussie-government-invented-wifi-and-sued-its-way-to-430-million.ars)

[11] "Australia's Biggest Patent Troll Goes After AT&T, Verizon and T-Mobile" (http://www.cbsnews.com/8301-505124_162-43340647/australias-biggest-patent-troll-goes-after-at038t-verizon-and-t-mobile/). *CBS News*. .

[12] Australian scientists cash in on Wi-Fi invention: SMH 1 April 2012 (http://www.smh.com.au/it-pro/government-it/australian-scientists-cash-in-on-wifi-invention-20120331-1w5gx.html)

[13] CSIRO wins legal battle over Wi-Fi patent: ABC 1 April 2012 (http://www.abc.net.au/news/2012-04-01/csiro-receives-multi-million-dollar-payment-for-wifi-technology/3925814)

[14] "Wi-Fi Alliance: Organization" (http://www.wi-fi.org/organization.php). *Official industry association web site*. . Retrieved August 23, 2011.

[15] US Patent and Trademark Office.

[16] "WiFi isn't short for "Wireless Fidelity"" (http://www.boingboing.net/2005/11/08/wifi_isnt_short_for_.html). boingboing.net. 2005-11-08. . Retrieved 2007-08-31.

[17] "Wireless Fidelity' Debunked" (http://www.wi-fiplanet.com/columns/article.php/3674591). Wi-Fi Planet. 2007-04-27. . Retrieved 2007-08-31.

[18] "What is the True Meaning of Wi-Fi?" (http://www.teleclick.ca/2005/12/what-is-the-true-meaning-of-wi-fi/). Teleclick. . Retrieved 2007-08-31.

[19] "Securing Wi-Fi Wireless Networks with Today's Technologies" (http://www.wi-fi.org/files/wp_4_Securing Wireless Networks_2-6-03.pdf). Wi-Fi Alliance. 2003-02-06. . Retrieved 2009-11-30.

[20] "WPA Deployment Guidelines for Public Access Wi-Fi Networks" (http://www.wi-fi.org/files/wp_6_WPA Deployment for Public Access_10-28-04.pdf). Wi-Fi Alliance. 2004-10-28. . Retrieved 2009-11-30.

[21] The Wi-Fi Alliance also developed technology that expanded the applicability of Wi-Fi, including a simple set up protocol (Wi-Fi Protected Set Up) and a peer to peer connectivity technology (Wi-Fi Peer to Peer) "Wi-Fi Alliance: Organization" (http://www.wi-fi.org/organization.php). www.wi-fi.org. . Retrieved 2009-10-22.

[22] "Wi-Fi Alliance: White Papers" (http://www.wi-fi.org/wp/wifi-alliance-certification/). www.wi-fi.org. . Retrieved 2009-10-22.

[23] "Wi-Fi Alliance: Programs" (http://www.wi-fi.org/certification_programs.php). www.wi-fi.org. . Retrieved 2009-10-22.

[24] "Mifi vs Joikuspot" (http://www.mificlub.com/2010/07/mifi-vs-joikuspot/). mificlub.com. . Retrieved 2010-10-09.

[25] The Telegraph - Say hello to India's first wirefree city (http://www.telegraphindia.com/1060820/asp/opinion/story_6632793.asp)

[26] "Sunnyvale Uses MetroFi" (http://www.unstrung.com/document.asp?doc_id=85119&WT.svl=wire1_1). unstrung.com. . Retrieved 2008-07-16.

[27] "Minneapolis moves ahead with wireless" (http://www.startribune.com/business/11286134.html). The Star Tribune. December 5, 2010. . Retrieved December 5, 2010.

[28] "London-wide wi-fi by 2012 pledge" (http://news.bbc.co.uk/2/hi/uk_news/england/london/8692103.stm). *BBC News*. 2010-05-19. . Retrieved 2010-05-19.

[29] "City of London Fires Up Europe's Most Advanced Wi-Fi Network" (http://www.govtech.com/dc/118717). www.govtech.com. . Retrieved 2007-05-14.

[30] "London gets a mile of free Wi-Fi" (http://www.zdnet.co.uk/news/networking/2005/04/18/london-gets-a-mile-of-free-wi-fi-39195421/). .zdnet.co.uk. . Retrieved 200-04-18.

[31] "Seoul Moves to Provide Free City-Wide WiFi Service" (http://blogs.voanews.com/breaking-news/2011/06/15/seoul-moves-to-provide-free-city-wide-wifi-service/). VOANEWS.COM. . Retrieved 1 April 2012.

[32] Deb Smit (October 5, 2011). "How Wi-Fi got its start on the campus of CMU, a true story" (http://popcitymedia.com/innovationnews/wifi100511.aspx). Pop City Media. . Retrieved October 6, 2011.

[33] "Wireless Andrew: Creating the World's First Wireless Campus" (http://www.cmu.edu/corporate/news/2007/features/wireless_andrew.shtml). Carnegie Mellon University. 2007. . Retrieved October 6, 2011.

[34] Wolter Lemstra; Vic Hayes; John Groenewegen (2010). *The innovation journey of Wi-Fi: the road to global success* (http://books.google.com/books?id=-OMoL5Irm08C&pg=PA121). Cambridge University Press. p. 121. ISBN 978-0-521-19971-1. . Retrieved October 6, 2011.

[35] "About the University" (http://www.drexel.edu/catalog/general/aboutuniversity.htm). Drexel.edu. . Retrieved 2011-10-14.

[36] "Wireless Home Networking with Virtual WiFi Hotspot" (http://techsansar.com/internetworking/wireless-home-networking-virtual-wifi-hotspot-2946/). Techsansar.com. 2011-01-24. . Retrieved 2011-10-14.

[37] "Wi-Fi Direct allows device-to-device links" (http://www.networkworld.com/news/2009/101409-wi-fi-direct.html?hpg1=bn). .

[38] "Wi-Fi gets personal: Groundbreaking Wi-Fi Direct launches today" (http://www.wi-fi.org/news_articles.php?f=media_news&news_id=1011). WiFi Alliance. 2010-10-25. . Retrieved 2011-01-15.

[39] "802.11n Delivers Better Range" (http://www.wi-fiplanet.com/tutorials/article.php/3680781). *Wi-Fi Planet*. 2007-05-31. .

[40] See for example IEEE Standard 802.15.4 section 1.2 scope

[41] "WiFi Mapping Software: Footprint" (http://www.alyrica.net/node/20). Alyrica Networks, Inc.. . Retrieved 2008-04-27.

[42] Wilson, Tracy V.. "How Municipal WiFi Works" (http://computer.howstuffworks.com/municipal-wifi.htm). computer.howstuffworks.com. . Retrieved 2008-03-12.

[43] "Apple.com Airport Utility Product Page" (http://www.apple.com/airportextreme/features/utility.html). Apple, Inc.. . Retrieved 2011-06-14.

[44] GainSpan specifically designs for Wi-Fi technology between Wi-Fi devices. Extremely useful. "GainSpan low-power, embedded Wi-Fi" (http://www.gainspan.com/technology/technology_overview.php). www.gainspan.com. . Retrieved 2010.

[45] "Ermanno Pietrosemoli has set a new record for the longest communication Wi-Fi link" (http://interred.wordpress.com/2007/06/18/ermanno-pietrosemoli-has-set-a-new-record-for-the-longest-communication-wi-fi-link/). . Retrieved 2008-03-10.

[46] "Wireless technology is irreplaceable for providing access in remote and scarcely populated regions" (http://www.apc.org/en/news/strategic/world/wireless-technology-irreplaceable-providing-access). . Retrieved 2008-03-10.

[47] "Long Distance WiFi Trial" (http://www.eslared.org.ve/articulos/Long Distance WiFi Trial.pdf) (PDF). . Retrieved 2008-03-10.

[48] "Quatech Rolls Out Airborne Embedded 802.11 Radio for M2M Market" (http://edageek.com/2008/04/18/embedded-wifi-radio/). . Retrieved 2008-04-29.

[49] "CIE article on embedded Wi-Fi for M2M applications" (http://www.cieonline.co.uk/cie2/articlen.asp?pid=1810&id=19742). . Retrieved 2008-08-27.

[50] "802.11 X Wireless Network in a Business Environment -- Pros and Cons." (http://networkbits.net/wireless-printing/80211-g-pros-cons-of-a-wireless-network-in-a-business-environment/). NetworkBits.net. . Retrieved 2008-04-08.

[51] Bernstein, Daniel J. (2002). "DNS forgery" (http://cr.yp.to/djbdns/forgery.html). . Retrieved 2010-03-24. "An attacker with access to your network can easily forge responses to your computer's DNS requests."

[52] Mateti, Prabhaker (2005). "Hacking Techniques in Wireless Networks" (http://www.cs.wright.edu/~pmateti/InternetSecurity/Lectures/WirelessHacks/Mateti-WirelessHacks.htm#_Toc77524658). Dayton, Ohio: Department of Computer Science and Engineering Wright State University. . Retrieved 2010-02-28.

[53] "Wireless Vulnerabilities & Exploits" (http://www.wirelessve.org/entries/show/WVE-2005-0020). wirelessve.org. . Retrieved 2008-04-15.

[54] "WPA2 Security Now Mandatory for Wi-Fi CERTIFIED Products" "WPA2 Security Now Mandatory for Wi-Fi CERTIFIED Products" (http://www.wi-fi.org/pressroom_overview.php?newsid=16). *Wi-Fi Alliance*. .

[55] http://www.kb.cert.org/vuls/id/723755 US CERT Vulnerability Note VU#723755

[56] "NoCat's goal is to bring you Infinite Bandwidth Everywhere for Free" (http://nocat.net/). Nocat.net. . Retrieved 2011-10-14.

[57] "Let's Warchalk" (http://www.blackbeltjones.com/warchalking/warchalking0_9.pdf) (PDF). Matt Jones. . Retrieved 2008-10-09.

[58] "Q&A: Wi-fi health concerns" (http://news.bbc.co.uk/2/hi/technology/6677051.stm). BBC News. 2007-05-21. . Retrieved 2011-10-14.

[59] "Electromagnetic Hypersensitivity (EMS)" (http://www.who.int/mediacentre/factsheets/fs296/en/), 2011

[60] "Official website" (http://www.globalnews.ca/programs/16x9/video.html?releasePID=ZDqUIbEn5ki6ptoRe__s2Oz5ODXbDiZ6). Globalnews.ca. . Retrieved 2011-10-14.

[61] ""Electromagnetic Hypersensitivity: A Systematic Review of Provocation Studies ", 2005" (http://www.psychosomaticmedicine.org/content/67/2/224.abstract). Psychosomaticmedicine.org. 2005-03-01. . Retrieved 2011-10-14.

[62] Avendaño C, Mata A, Juarez Villanueva AM, Martinez VS, Sanchez Sarmiento CA (2010) "Laptop expositions affect motility and induce DNA fragmentation in human spermatozoa in vitro by a non-thermal effect: a preliminary report" *American Society for Reproductive Medicine*, 66th Annual Meeting: O-249.

[63] Papageorgiou CC, Hountala CD, Maganioti AE, Kyprianou MA, Rabavilas AD, Papadimitriou GN, Capsalis CN (2011) "Effects of wi-fi signals on the p300 component of event-related potentials during an auditory hayling task" *J Integr Neurosci.* **10(2)**: 189-202; PMID 21714138.

References

Further reading

- *Wireless Networking in the Developing World* (PDF book)

External links

- The Wi-Fi Alliance (http://wi-fi.org/)
- List of Wi-Fi certified products (http://www.wi-fi.org/search_products.php?search=1&advanced=1&lang=en&filter_company_id=&filter_category_id=&filter_subcategory=&filter_cid=&date_from=&date_to=&selected_certifications[]=50&x=44&y=15/)

Symbian

Symbian

Company / developer	Accenture on behalf of Nokia[1]
Programmed in	C++[2]
OS family	Embedded operating system
Working state	Current (will get support until at least 2016)
Source model	Proprietary[3]
Initial release	1997 as EPOC32[4]
Latest stable release	Nokia Belle Feature Pack 1 (Updated Symbian^3)
Latest unstable release	Nokia Belle Feature Pack 2
Marketing target	Nokia line of Smartphones
Supported platforms	ARM, x86[5]
Kernel type	Real Time Microkernel
Default user interface	Avkon[6] (Graphical)
License	Proprietary
Official website	worldwide.nokia.com/belle/ [7]

Symbian is a mobile operating system (OS) and computing platform designed for smartphones and currently maintained by Accenture.[8] The Symbian platform is the successor to Symbian OS and Nokia Series 60; unlike Symbian OS, which needed an additional user interface system, Symbian includes a user interface component based on S60 5th Edition. The latest version, Symbian^3, was officially released in Q4 2010, first used in the Nokia N8. In May 2011 an update, Symbian Anna, was officially announced, followed by Nokia Belle (previously Symbian Belle) in August 2011.[9][10]

Symbian OS was originally developed by Symbian Ltd.[11] It is a descendant of Psion's EPOC and runs exclusively on ARM processors, although an unreleased x86 port existed.

Some estimates indicate that the number of mobile devices shipped with the Symbian OS up to the end of Q2 2010 is 385 million.[12]

By 5 April 2011, Nokia released Symbian under a new license and converted to a proprietary model as opposed to an open source project.[3]

On 11 February 2011, Nokia announced that it would migrate from Symbian to Windows Phone 7. Nokia CEO Stephen Elop announced Nokia's first Windows phones at Nokia World 2011: the Lumia 800 and Lumia 710. These phones were launched on 14 November 2011.[13] On 22 June 2011 Nokia made an agreement with Accenture for an outsourcing program. Accenture will provide Symbian-based software development and support services to Nokia through 2016; about 2,800 Nokia employees became Accenture employees as of October 2011.[14] The transfer was completed on 30 September 2011.[8]

History

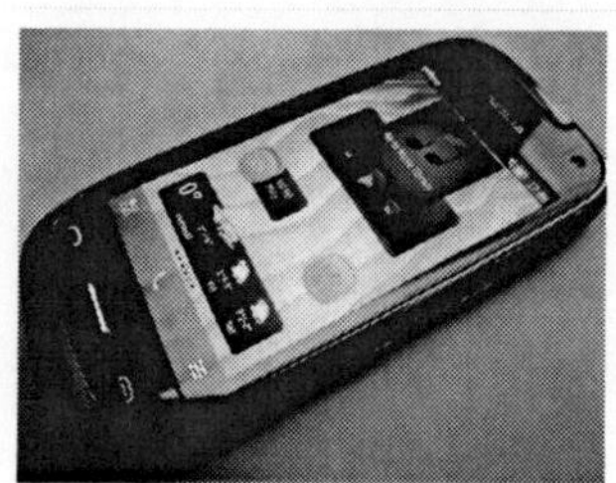
The Nokia C7 running the Nokia Belle OS

The Symbian platform was created by merging and integrating software assets contributed by Nokia, NTT DoCoMo, Sony Ericsson and Symbian Ltd., including Symbian OS assets at its core, the S60 platform, and parts of the UIQ and MOAP(S) user interfaces.

In December 2008, Nokia bought Symbian Ltd., the company behind Symbian OS; consequently, Nokia became the major contributor to Symbian's code, since it then possessed the development resources for both the Symbian OS core and the user interface. Since then Nokia has been maintaining its own code repository for the platform development, regularly releasing its development to the public repository.[15] Symbian was intended to be developed by a community led by the Symbian Foundation, which was first announced in June 2008 and which officially launched in April 2009. Its objective was to publish the source code for the entire Symbian platform under the OSI- and FSF-approved Eclipse Public License (EPL). The code was published under EPL on 4 February 2010; Symbian Foundation reported this event to be the largest codebase transitioned to Open Source in history.[16][17]

However, some important components within Symbian OS were licensed from third parties, which prevented the foundation from publishing the full source under EPL immediately; instead much of the source was published under a more restrictive Symbian Foundation License (SFL) and access to the full source code was limited to member companies only, although membership was open to any organisation.[18]

In November 2010, the Symbian Foundation announced that due to a lack of support from funding members, it would transition to a licensing-only organisation; Nokia announced it would take over the stewardship of the Symbian platform. Symbian Foundation will remain the trademark holder and licensing entity and will only have non-executive directors involved.

On 11 February 2011, Nokia announced a partnership with Microsoft that would see it adopt Windows Phone 7 for smartphones, reducing the number of devices running Symbian over the coming two years.[13] As a consequence, the use of the Symbian platform for building mobile applications dropped rapidly. Research in June 2011 indicated that over 39% of mobile developers using Symbian at the time of publication were planning to abandon the platform.[19]

By 5 April 2011, Nokia ceased to open source any portion of the Symbian software and reduced its collaboration to a small group of pre-selected partners in Japan.[3] Source code released under the EPL remains available in third party repositories.[20][21]

Features

User interface

Symbian has had a native graphics toolkit since its inception, known as AVKON (formerly known as Series 60). S60 was designed to be manipulated by a keyboard-like interface metaphor, such as the ~15-key augmented telephone keypad, or the mini-QWERTY keyboards. AVKON-based software is binary-compatible with Symbian versions up to and including Symbian^3.

Symbian^3 includes the Qt framework, which is now the recommended user interface toolkit for new applications. Qt can also be installed on older Symbian devices.

Symbian^4 was planned to introduce a new GUI library framework specifically designed for a touch-based interface, known as "UI Extensions for Mobile" or UIEMO (internal project name "Orbit"), which was built on top of Qt Widget; a preview was released in January 2010, however in October 2010 Nokia announced that Orbit/UIEMO has been cancelled.

Nokia currently recommends that developers use Qt Quick with QML, the new high-level declarative UI and scripting framework for creating visually rich touchscreen interfaces that allows development for both Symbian and MeeGo; it will be delivered to existing Symbian^3 devices as a Qt update. When more applications gradually feature a user interface reworked in Qt, the legacy S60 framework (AVKON) will be deprecated and no longer included with new devices at some point, thus breaking binary compatibility with older S60 applications.[22][23]

Browser

Symbian^3 and earlier have a native WebKit based browser; indeed, Symbian was the first mobile platform to make use of WebKit (in June 2005).[24] Some older Symbian models have Opera Mobile as their default browser.

Nokia released a new browser with the release of Symbian Anna with improved speed and an improved user interface.[25]

Multiple language support

Symbian has strong localization support enabling manufacturers and 3rd party application developers to localize their Symbian based products in order to support global distribution.

Current Symbian release (Symbian Belle) has support for 48 languages, which Nokia makes available on device in language packs (set of languages which cover the languages commonly spoken in the area where the device variant is intended to be sold). All language packs have in common English (or a locally relevant dialect of it).

The supported languages [with dialects] (and scripts) in Symbian Belle are:

- Arabic (Arabic),
- Basque (Latin),
- Bulgarian (Cyrillic),
- Catalan (Latin),
- Chinese [PRC] (Simplified Chinese),
- Chinese [Hong Kong] (Traditional Chinese),
- Chinese [Taiwan] (Traditional Chinese),
- Croatian (Latin),
- Czech (Latin),
- Danish (Latin),
- Dutch (Latin),
- English [UK] (Latin),
- English [US] (Latin),
- Estonian (Latin),
- Finnish (Latin),
- French (Latin),
- French [Canadian] (Latin),
- Galician (Latin),
- German (Latin),
- Greek (Greek),
- Hebrew (Hebrew),
- Hindi (Hi),
- Hungarian (Latin),
- Icelandic (Latin),
- Indonesian [Bahasa Indonesia] (Latin),
- Italian (Latin),
- Kazakh (Cyrillic),
- Latvian (Latin),
- Lithuanian (Latin),
- Malay [Bahasa Malaysia] (Latin),
- Norwegian (Latin),
- Persian [Farsi],
- Polish (Latin),
- Portuguese (Latin),
- Portuguese [Brazilian] (Latin),
- Romanian [Romania] (Latin),
- Russian (Cyrillic),
- Serbian (Latin),
- Slovak (Latin),
- Slovene (Latin),
- Spanish (Latin),
- Spanish [Latin America] (Latin),
- Swedish (Latin),
- Tagalog [Filipino] (Latin),
- Thai (Thai),
- Turkish (Latin),
- Ukrainian (Cyrillic),
- Urdu (Arabic),
- Vietnamese (Latin).
- marathi(maharashtra)

Symbian Belle marks the introduction of Kazakh, while Japanese and Korean is no longer supported.

Application development

From 2010, Symbian switched to using standard C++ with Qt as the main SDK, which can be used with either Qt Creator or Carbide.c++. Qt supports the older Symbian/S60 3rd (starting with Feature Pack 1, aka S60 3.1) and Symbian/S60 5th Edition (aka S60 5.0) releases, as well as the new Symbian platform. It also supports Maemo and MeeGo, Windows, Linux and Mac OS X.[26][27]

Alternative application development can be done using Python (see Python for S60), Adobe Flash Lite or Java ME.

Symbian OS previously used a Symbian specific C++ version, along with Carbide.c++ integrated development environment (IDE), as the native application development environment.

Web Run time (WRT) is a portable application framework that allows creating widgets on the S60 Platform; it is an extension to the S60 WebKit based browser that allows launching multiple browser instances as separate JavaScript applications.[28][29]

Application development

Qt

As of 2010, the SDK for Symbian is standard C++, using Qt. It can be used with either Qt Creator, or Carbide (the older IDE previously used for Symbian development).[26][30] A phone simulator allows testing of Qt apps. Apps compiled for the simulator are compiled to native code for the development platform, rather than having to be emulated.[31] Application development can either use C++ or QML.

Symbian C++

As Symbian OS is written in C++ using Symbian Software's coding standards, it is naturally possible to develop using Symbian C++, although it is not a standard implementation. Before the release of the Qt SDK, this was the standard development environment. There were multiple platforms based on Symbian OS that provided software development kit (SDKs) for application developers wishing to target Symbian OS devices, the main ones being UIQ and S60. Individual phone products, or families, often had SDKs or SDK extensions downloadable from the maker's website too.

The SDKs contain documentation, the header files and library files needed to build Symbian OS software, and a Windows-based emulator ("WINS"). Up until Symbian OS version 8, the SDKs also included a version of the GNU Compiler Collection (GCC) compiler (a cross-compiler) needed to build software to work on the device.

Symbian OS 9 and the Symbian platform use a new application binary interface (ABI) and needed a different compiler. A choice of compilers is available including a newer version of GCC (see external links below).

Unfortunately, Symbian C++ programming has a steep learning curve, as Symbian C++ requires the use of special techniques such as descriptors, active objects and the cleanup stack. This can make even relatively simple programs initially harder to implement than in other environments. It is possible that the techniques, developed for the much more restricted mobile hardware and compilers of the 1990s, caused extra complexity in source code because programmers are required to concentrate on low-level details instead of more application-specific features. As of 2010, these issues are no longer the case when using standard C++, with the Qt SDK.

Symbian C++ programming is commonly done with an integrated development environment (IDE). For earlier versions of Symbian OS, the commercial IDE CodeWarrior for Symbian OS was favoured. The CodeWarrior tools were replaced during 2006 by Carbide.c++, an Eclipse-based IDE developed by Nokia. Carbide.c++ is offered in four different versions: Express, Developer, Professional, and OEM, with increasing levels of capability. Fully featured software can be created and released with the Express edition, which is free. Features such as UI design, crash debugging etc. are available in the other, charged-for, editions. Microsoft Visual Studio 2003 and 2005 are also supported via the Carbide.vs plugin.

Other languages

Symbian devices can also be programmed using Python, Java ME, Flash Lite, Ruby, .NET, Web Runtime (WRT) Widgets and Standard C/C++.[32]

Visual Basic programmers can use NS Basic to develop apps for S60 3rd Edition and UIQ 3 devices.

In the past, Visual Basic, Visual Basic .NET, and C# development for Symbian were possible through AppForge Crossfire, a plugin for Microsoft Visual Studio. On 13 March 2007 AppForge ceased operations; Oracle purchased the intellectual property, but announced [33] that they did not plan to sell or provide support for former AppForge products. Net60 [34], a .NET compact framework for Symbian, which is developed by redFIVElabs, is sold as a commercial product. With Net60, VB.NET and C# (and other) source code is compiled into an intermediate language (IL) which is executed within the Symbian OS using a just-in-time compiler. (As of 18/1/10 RedFiveLabs has ceased development of Net60 with this announcement on their landing page: "At this stage we are pursuing some options to sell the IP so that Net60 may continue to have a future".)

There is also a version of a Borland IDE for Symbian OS. Symbian OS development is also possible on Linux and Mac OS X using tools and methods developed by the community, partly enabled by Symbian releasing the source code for key tools. A plugin that allows development of Symbian OS applications in Apple's Xcode IDE for Mac OS X was available.[35]

Java ME applications for Symbian OS are developed using standard techniques and tools such as the Sun Java Wireless Toolkit (formerly the J2ME Wireless Toolkit). They are packaged as JAR (and possibly JAD) files. Both CLDC and CDC applications can be created with NetBeans. Other tools include SuperWaba, which can be used to build Symbian 7.0 and 7.0s programs using Java.

Nokia S60 phones can also run Python scripts when the interpreter Python for S60 is installed, with a custom made API that allows for Bluetooth support and such. There is also an interactive console to allow the user to write Python scripts directly from the phone.

Deployment

Once developed, Symbian applications need to find a route to customers' mobile phones. They are packaged in SIS files which may be installed over-the-air, via PC connect, Bluetooth or on a memory card. An alternative is to partner with a phone manufacturer and have the software included on the phone itself. Applications must be Symbian Signed [36] for Symbian OS 9.x in order to make use of certain capabilities (system capabilities, restricted capabilities and device manufacturer capabilities).[37] Applications can now be signed for free.[38]

Architecture

Technology domains and packages

Symbian's design is subdivided into **technology domains**,[39] each of which comprises a number of software **packages**.[40] Each technology domain has its own roadmap, and the Symbian Foundation has a team of technology managers who manage these technology domain roadmaps.

Every package is allocated to exactly one technology domain, based on the general functional area to which the package contributes and by which it may be influenced. By grouping related packages by themes, the Symbian Foundation hopes to encourage a strong community to form around them and to generate discussion and review.

The Symbian System Model[41] illustrates the scope of each of the technology domains across the platform packages.

Packages are owned and maintained by a package owner, a named individual from an organization member of the Symbian Foundation, who accepts code contributions from the wider Symbian community and is responsible for package.

Symbian kernel

The Symbian kernel (EKA2) supports sufficiently fast real-time response to build a single-core phone around it—that is, a phone in which a single processor core executes both the user applications and the signalling stack.[42] The real-time kernel has a microkernel architecture containing only the minimum, most basic primitives and functionality, for maximum robustness, availability and responsiveness. It has been termed a nanokernel, because it needs an extended kernel to implement any other abstractions. It contains a scheduler, memory management and device drivers, with networking, telephony and file system support services in the OS Services Layer or the Base Services Layer. The inclusion of device drivers means the kernel is not a *true* microkernel.

Design

Symbian features pre-emptive multitasking and memory protection, like other operating systems (especially those created for use on desktop computers). EPOC's approach to multitasking was inspired by VMS and is based on asynchronous server-based events.

Symbian OS was created with three systems design principles in mind:

1. the integrity and security of user data is paramount
2. user time must not be wasted
3. all resources are scarce

To best follow these principles, Symbian uses a microkernel, has a request-and-callback approach to services, and maintains separation between user interface and engine. The OS is optimised for low-power battery-based devices and for ROM-based systems (e.g. features like XIP and re-entrancy in shared libraries). Applications, and the OS itself, follow an object-oriented design: Model-view-controller (MVC).

Later OS iterations diluted this approach in response to market demands, notably with the introduction of a real-time kernel and a platform security model in versions 8 and 9.

There is a strong emphasis on conserving resources which is exemplified by Symbian-specific programming idioms like descriptors and a cleanup stack. Similar methods exist to conserve disk space, though disks on Symbian devices are usually flash memory. Further, all Symbian programming is event-based, and the central processing unit (CPU) is switched into a low power mode when applications are not directly dealing with an event. This is done via a programming idiom called active objects. Similarly the Symbian approach to threads and processes is driven by reducing overheads.

Operating system

The All over Model contains the following layers, from top to bottom:

- UI Framework Layer
- Application Services Layer
  - Java ME
- OS Services Layer
  - generic OS services
  - communications services
  - multimedia and graphics services
  - connectivity services
- Base Services Layer
- Kernel Services & Hardware Interface Layer

The Base Services Layer is the lowest level reachable by user-side operations; it includes the File Server and User Library, a Plug-In Framework which manages all plug-ins, Store, Central Repository, DBMS and cryptographic services. It also includes the Text Window Server and the Text Shell: the two basic services from which a completely functional port can be created without the need for any higher layer services.

Symbian has a microkernel architecture, which means that the minimum necessary is within the kernel to maximise robustness, availability and responsiveness. It contains a scheduler, memory management and device drivers, but other services like networking, telephony and filesystem support are placed in the OS Services Layer or the Base Services Layer. The inclusion of device drivers means the kernel is not a *true* microkernel. The EKA2 real-time kernel, which has been termed a nanokernel, contains only the most basic primitives and requires an extended kernel to implement any other abstractions.

Symbian is designed to emphasise compatibility with other devices, especially removable media file systems. Early development of EPOC led to adopting FAT as the internal file system, and this remains, but an object-oriented persistence model was placed over the underlying FAT to provide a POSIX-style interface and a streaming model. The internal data formats rely on using the same APIs that create the data to run all file manipulations. This has resulted in data-dependence and associated difficulties with changes and data migration.

There is a large networking and communication subsystem, which has three main servers called: ETEL (EPOC telephony), ESOCK (EPOC sockets) and C32 (responsible for serial communication). Each of these has a plug-in scheme. For example, ESOCK allows different ".PRT" protocol modules to implement various networking protocol schemes. The subsystem also contains code that supports short-range communication links, such as Bluetooth, IrDA and USB.

There is also a large volume of user interface (UI) Code. Only the base classes and substructure were contained in Symbian OS, while most of the actual user interfaces were maintained by third parties. This is no longer the case. The three major UIs — S60, UIQ and MOAP — were contributed to Symbian in 2009. Symbian also contains graphics, text layout and font rendering libraries.

All native Symbian C++ applications are built up from three framework classes defined by the application architecture: an application class, a document class and an application user interface class. These classes create the fundamental application behaviour. The remaining needed functions, the application view, data model and data interface, are created independently and interact solely through their APIs with the other classes.

Many other things do not yet fit into this model — for example, SyncML, Java ME providing another set of APIs on top of most of the OS and multimedia. Many of these are frameworks, and vendors are expected to supply plug-ins to these frameworks from third parties (for example, Helix Player for multimedia codecs). This has the advantage that the APIs to such areas of functionality are the same on many phone models, and that vendors get a lot of flexibility. But it means that phone vendors needed to do a great deal of integration work to make a Symbian OS phone.

Symbian includes a reference user-interface called "TechView." It provides a basis for starting customisation and is the environment in which much Symbian test and example code runs. It is very similar to the user interface from the Psion Series 5 personal organiser and is not used for any production phone user interface.

Devices and feature comparison

On 16 November 2006, the 100 millionth smartphone running the OS was shipped.[43] As of 21 July 2009, more than 250 million devices running Symbian OS had been shipped.[44]

- The Nokia S60 interface is used in various phones, the first being the Nokia 7650. The Nokia N-Gage and Nokia N-Gage QD gaming/smartphone combos are also S60 platform devices. It was also used on other manufacturers' phones such as the Siemens SX1 and Samsung SGH-Z600. Recently, more advanced devices using S60 include the Nokia 6xxx, the Nseries (except Nokia N8xx and N9xx), the Eseries and some models of the Nokia XpressMusic mobiles.
- Fujitsu, Mitsubishi, Sony Ericsson and Sharp developed phones for NTT DoCoMo in Japan, using an interface developed specifically for DoCoMo's FOMA "Freedom of Mobile Access" network brand. This UI platform is

called MOAP "Mobile Oriented Applications Platform" and is based on the UI from earlier Fujitsu FOMA models. The user cannot install new C++ applications.

User interfaces that run on or are based on Symbian OS include:

- S60, formerly Series 60, used by Nokia and others
- Series 80, previously used by Nokia
- Series 90, previously used by Nokia
- UIQ, previously used by Sony-Ericsson and many other manufacturers
- MOAP, Mobile Oriented Applications Platform, used by NTT DoCoMo's FOMA service
- OPP, successor of MOAP, used on NTT DoCoMo's FOMA phone

Versions that are actively marketed as of September 2011 are Symbian^3 (and its updated Symbian Anna and Nokia Belle variants), Symbian^2, Symbian^1 (previously known as Series 60 5th Edition), and Series 60 3rd Edition Feature Pack 2. For features of older versions, see history of Symbian. Note that the operating system supporting a certain feature does not imply that all devices running on it have that feature available, especially if it involves expensive hardware, such as HDMI output.

Feature	Symbian^3/Anna/Belle	Symbian^2[45]	Symbian^1/Series 60 5th Edition	Series 60 3rd Edition Feature Pack 2	Series 80
Year released	2010 (Symbian^3), 2011 (Symbian Anna, Nokia Belle)	2010	2008	2008	2002
Company	Symbian Foundation	Symbian Foundation	Symbian Foundation	Symbian Foundation	Symbian Foundation
Symbian OS version	9.5 (Symbian^3/Symbian Anna), 10.1 (Nokia Belle)		9.4	9.3	
Series 60 version	5.2 (Symbian^3/Symbian Anna)[46], 5.3 (Nokia Belle), 5.4 (Nokia Belle FP1)	5.1	5th Edition	3rd Edition Feature Pack 2	N/A
Touch input support	Yes	Yes	Yes	No	No
Multi touch input support	Yes		No	No	No
Number of customizable home screens	Three to six (Five on Nokia E6, six on Nokia Belle)		One	Two	One
Wi-Fi version support	B, G, N		B, G	B, G	B, G
USB on the go support	Yes		No	No	
DVB-H support	Yes, with extra headset[47]	Unknown, but have 1seg support[48]	Yes, with extra headset	Yes, with extra headset	
Short range FM transmitter support	Yes		Yes	Yes	No
FM radio support	Yes		Yes	Yes	No

Feature	Symbian^3/Anna/Belle	Symbian^2	Symbian^1/Series 60 5th Edition	Series 60 3rd Edition Feature Pack 2	Series 80
Adobe Flash support	Yes, Flash Lite native version 4.0, upgradable		Yes, Flash Lite native version 3.1, upgradable	Yes, Flash Lite native version 3.1, upgradable	No
Microsoft Silverlight support	No[49]		Yes[50][51]	No[52]	No
OpenGL ES support	Yes, version 2.0				No
SQLite support	Yes		Yes	Yes[53]	
CPU architecture support	ARM	SH-Mobile	ARM	ARM	ARM
Programmed in	C++, Qt		C++, Qt	C++, Qt	
License	Eclipse Public License; Since 31 March 2011: Nokia Symbian License 1.0	proprietary SFL license, while some portions of source code are EPL licensed.			
Public issues list	No more				
Package manager	.sis, .sisx		.sis, .sisx	.sis, .sisx	.sis, .sisx
Non English languages support	Yes	mainly Japanese,	Yes	Yes	Yes
Underlining spell checker	Yes	Yes[54]	Yes	Yes	
Keeps state on shutdown or crash	No		No	No	No
Internal search	Yes	Yes[48]	Yes	Yes	Yes
Proxy server	Yes		Yes	Yes	Yes
On-device encryption	Yes	Yes[48]	Yes	Yes	
Cut, copy, and paste support	Yes	Yes[54]	Yes	Yes	Yes
Undo	No		No	No	Yes
Default Web Browser for S60, WebKit engine	version 7.2, engine version 525 (Symbian^3);[55] version 7.3, engine version 533.4 (Symbian Anna)		version 7.1.4, engine version 525; version 7.3, engine version 533.4 (for 9 selected units after firmware updates released in summer 2011)	engine version 413 (Nokia N79)	N/A

third-party software store	Nokia store	i-αppli/i-Widget[54]	Nokia store	Nokia store	
Email sync protocol support	POP3, IMAP	i-mode mail[54]	POP3, IMAP	POP3, IMAP	POP3, IMAP
Feature	**Symbian^3/Anna/Belle**	**Symbian^2**	**Symbian^1/Series 60 5th Edition**	**Series 60 3rd Edition Feature Pack 2**	**Series 80**
Push alerts	Yes		Yes	Yes	Yes
Voice recognition	Yes	Yes	Yes	Yes	
Tethering	USB, Bluetooth; mobile Wi-Fi hotspot, with third-party software		USB, Bluetooth; mobile Wi-Fi hotspot, with third-party software	USB, Bluetooth; mobile Wi-Fi hotspot, with third-party software	USB, Bluetooth;
Text, document support	Mobile Office Applications, PDF	Mobile Office Applications, PDF	Mobile Office Applications, PDF	Mobile Office Applications, PDF	Mobile Office Applications, PDF
Audio playback	All	wma,[48] aac	All	All	wav, mp3
Video playback	H.263, H.264, WMV, MPEG4, MPEG4@ HD 720p 25–30 frame/s, MKV, DivX, XviD	WMV,[54] MPEG4	H.263, WMV, MPEG4, 3GPP, 3GPP2	H.263, WMV, MPEG4, 3GPP, 3GPP2	H.263, 3GPP, 3GPP2
Turn-by-turn GPS	Yes, with third-party software, or free global Nokia Nokia Maps which works offline	Yes, with monthly paid Docomo Map Navi[56] (□ □[57])	Yes, with third-party software, or free global Nokia Nokia Maps which works offline	Yes, with third-party software, or free global Nokia Nokia Maps which works offline	Yes, with third-party software
Video out	Nokia AV, PAL, NTSC, HDMI	HDMI, and	Nokia AV, PAL, NTSC	Nokia AV, PAL, NTSC	No
Multitasking	Yes		Yes	Yes	Yes
Desktop interactive widgets	Yes	Yes	Yes	No	
Integrated hardware keyboard	Yes	Yes	Yes	Yes	Yes
Bluetooth keyboard	Yes	Yes[48]	Yes	Yes	Yes
Video conference front video camera	Yes	Yes	Yes	Yes	Yes
Can share data via Bluetooth with all devices	Yes	Yes	Yes	Yes	Yes
Skype, third-party software	Yes[58]		Yes[58]	Yes[58]	

Facebook IM chat	Yes		Yes	Yes	
Secure Shell (SSH)	Yes, third-party software		Yes, third-party software	Yes, third-party software	
OpenVPN	No, Nokia VPN can be used		No, Nokia VPN can be used	No, Nokia VPN can be used	Yes, third-party software
Remote frame buffer	?				
Screenshot	Yes, third-party software[59]		Yes, third-party software[59]	Yes, third-party software[59]	Yes
GPU acceleration	Yes				No
Official SDK platform(s)	Cross-platform, Windows (preferred is Qt), Carbide.c++, Java ME, Web Runtime Widgets (WRT), Flash lite, Python for Symbian		Cross-platform, Windows (preferred is Qt), Carbide.c++, Java ME, Web Runtime Widgets (WRT), Flash lite, Python for Symbian	Cross-platform, Windows (preferred is Qt), Carbide.c++, Java ME, Web Runtime Widget (WRT), Flash lite, Python for Symbian	Cross-platform, Windows (preferred is Qt), Carbide.c++, Java ME, third-party software (OPL)
Feature	**Symbian^3/Anna/Belle**	**Symbian^2**	**Symbian^1/Series 60 5th Edition**	**Series 60 3rd Edition Feature Pack 2**	**Series 80**
First device(s)	Nokia N8 (Symbian^3), Nokia C7 (Symbian^3), Nokia X7, Nokia E6 (Anna), Nokia 603, Nokia 700, Nokia 701 (Belle)	NTT DOCOMO STYLE Series F-07B	Nokia 5800 (2 October 2008)	Nokia N96, Nokia N78, Nokia 6210 Navigator and Nokia 6220 Classic (11 February 2008)	Nokia 9210
Devices	Nokia N8, Nokia C6-01, Nokia C7-00, Nokia E7-00, Nokia E6, Nokia X7, Nokia 500, Nokia 603, Nokia 700, Nokia 701, Nokia 808 PureView	NTT DoCoMo: F-06B*,[60] F-07B*,[60] F-08B*,[60] SH-07B†,[60] F-10B,[61] Raku-Raku Phone 7,[61] F-01C*,[62] F-02C*,[62] F-03C*,[62] F-04C*,[62] F-05C*,[62] SH-01C†,[62] SH-02C†,[62] SH-04C†,[62] SH-05C†,[62] SH-06C†,[62] Touch Wood SH-08C†[62]	**Nokia:** 5228, 5230, 5233, 5235, 5250, 5530 XpressMusic, 5800 XpressMusic, 5800 Navigation Edition, C5-03, C6-00, N97, N97 mini, X6; **Samsung:** i8910 Omnia HD,[63] **Sony Ericsson:** Satio, Vivaz, Vivaz Pro	**Nokia:** 5320 XpressMusic, 5630 XpressMusic, 5730 XpressMusic, 6210 Navigator, 6220 Classic, 6650 fold, 6710 Navigator, 6720 Classic, 6730 Classic, 6760 Slide, 6790 Surge, E5-00, E52, E55, E71, E72, E75, N78, N79, N85, N86 8MP, N96, X5, C5-00; **Samsung:** GT-i8510 (INNOV8), GT-I7110, SGH-L870	Nokia 9210, Nokia 9300, Nokia 9300i, Nokia 9500
Feature	**Symbian^3/Anna/Belle**	**Symbian^2**	**Symbian^1/Series 60 5th Edition**	**Series 60 3rd Edition Feature Pack 2**	**Series 80**

* manufactured by Fujitsu

† manufactured by Sharp

Market share and competition

In the number of "smart mobile device" sales, Symbian devices were the market leaders for 2010. Statistics showed that Symbian devices formed a 37.6% share of smart mobile devices sold, with Android having 22.7%, RIM having 16%, and Apple having 15.7% (via iOS).[64]

Prior reports on device shipments as published in February 2010 showed that the Symbian devices formed a 47.2% share of the smart mobile devices shipped in 2009, with RIM having 20.8%, Apple having 15.1% (via iOS), Microsoft having 8.8% (via Windows CE and Windows Mobile) and Android having 4.7%.[65] Other competitors include webOS, Qualcomm's BREW, SavaJe, Linux and MontaVista Software.

Symbian has lost market share over the years as the market has dramatically grown, with new competing platforms entering the market, though its sales have increased during the same timeframe. E.g., although Symbian's share of the global smartphone market dropped from 52.4% in 2008 to 47.2% in 2009, shipments of Symbian devices grew 4.8%, from 74.9 million units to 78.5 million units.[65] From Q2 2009 to Q2 2010, shipments of Symbian devices grew 41.5%, by 8.0 million units, from 19,178,910 units to 27,129,340; compared to an increase of 9.6 million units for Android, 3.3 million units for RIM, and 3.2 million units for Apple.[66] In 2006, Symbian had 73% of the smartphone market,[67] compared with 22.1% of the market in the second quarter of 2011.[68] Over the course of 2009–2011, Nokia, Motorola, Samsung, LG, and Sony Ericsson announced their withdrawal from Symbian in favour of alternative platforms including Google's Android, Microsoft's Windows Phone, and Samsung's bada.[69][70][71][72]

Criticisms

The users of Symbian in the countries with non-Latin alphabets (such as Russia, Ukraine and others) have been criticizing the complicated method of language switching for many years.[73] For example, if a user wants to type a Latin letter, he must call the menu, click the languages item, use arrow keys to choose, for example, the English language from among many other languages, and then press the 'OK' button. After typing the Latin letter, the user must repeat the procedure to return to his native keyboard. This method slows down typing significantly. In touch-phones and QWERTY phones the procedure is slightly different but remains time-consuming. All other mobile operating systems, as well as Nokia's S40 phones, enable switching between two initially selected languages by one click or a single gesture.

Early versions of the firmware for the original Nokia N97, running on Symbian^1/Series 60 5th Edition have been heavily criticized.

In November 2010, Smartphone blog *All About Symbian* criticized the performance of Symbian's default web browser and recommended the alternative browser Opera Mobile.[74] Nokia's Senior Vice President Jo Harlow promised an updated browser in the first quarter of 2011.[75]

Malware

Symbian OS is subject to a variety of viruses, the best known of which is Cabir. Usually these send themselves from phone to phone by Bluetooth. So far, none have taken advantage of any flaws in Symbian OS – instead, they have all asked the user whether they would like to install the software, with somewhat prominent warnings that it can't be trusted, although some rely on social engineering, often in the form of messages that come with the malware, purporting to be a utility, game or some other application for Symbian.

However, with a view that the average mobile phone user shouldn't have to worry about security, Symbian OS 9.x adopted a UNIX-style capability model (permissions per process, not per object). Installed software is theoretically unable to do damaging things (such as costing the user money by sending network data) without being digitally signed – thus making it traceable. Commercial developers who can afford the cost can apply to have their software signed via the Symbian Signed [36] program. Developers also have the option of self-signing their programs. However, the set of available features does not include access to Bluetooth, IrDA, GSM CellID, voice calls, GPS and few others. Some operators have opted to disable all certificates other than the Symbian Signed certificates.

Some other hostile programs are listed below, but all of them still require the input of the user to run.

- Drever.A is a malicious SIS file trojan that attempts to disable the automatic startup from Simworks and Kaspersky Symbian Anti-Virus applications.
- Locknut.B is a malicious SIS file trojan that pretends to be a patch for Symbian S60 [76] mobile phones. When installed, it drops a binary that will crash a critical system service component. This will prevent any application from being launched in the phone.
- Mabir.A is basically Cabir with added MMS functionality. The two are written by the same author, and the code shares many similarities. It spreads using Bluetooth via the same routine as early variants of Cabir. As Mabir.A activates, it will search for the first phone it finds, and starts sending copies of itself to that phone.
- Fontal.A is an SIS file trojan that installs a corrupted file which causes the phone to fail at reboot. If the user tries to reboot the infected phone, it will be permanently stick on the reboot, and cannot be used without disinfection – that is, the use of the reformat key combination which causes the phone to lose all data. Being a trojan, Frontal cannot spread by itself – the most likely way for the user to get infected would be to acquire the file from untrusted sources, and then install it to the phone, inadvertently or otherwise.

A new form of malware threat to Symbian OS in the form of 'cooked firmware' was demonstrated at the International Malware Conference, MalCon, December 2010, by Indian hacker Atul Alex.[77][78]

Bypassing platform security

Symbian OS 9.x devices can be hacked to remove the platform security introduced in OS 9.1 onwards, allowing users to execute unsigned code.[79] This allows altering system files, and access to previously locked areas of the OS. The hack was criticised by Nokia for potentially increasing the threat posed by mobile viruses as unsigned code can be executed.[80]

Version history

Version	Description
EPOC16	EPOC16, originally simply named EPOC, was the operating system developed by Psion in the late 1980s and early 1990s for Psion's "SIBO" (SIxteen Bit Organisers) devices. All EPOC16 devices featured an 8086-family processor and a 16-bit architecture. EPOC16 was a single-user preemptive multitasking operating system, written in Intel 8086 assembler language and C and designed to be delivered in ROM. It supported a simple programming language called Open Programming Language (OPL) and an integrated development environment (IDE) called OVAL. SIBO devices included the: MC200, MC400, Series 3 (1991–98), Series 3a, Series 3c, Series 3mx, Siena, Workabout and Workabout mx. The MC400 and MC200, the first EPOC16 devices, shipped in 1989. EPOC16 featured a primarily 1-bit-per-pixel, keyboard-operated graphical interface[81] — the hardware for which it was designed did not have pointer input. In the late 1990s, the operating system was referred to as **EPOC16** to distinguish it from Psion's then-new EPOC32 OS.
EPOC32 (releases 1 to 5)	The first version of EPOC32, Release 1 appeared on the Psion Series 5 ROM v1.0 in 1997. Later, ROM v1.1 featured Release 3 (Release 2 was never publicly available.) These were followed by the Psion Series 5mx, Revo / Revo plus, Psion Series 7 / netBook and netPad (which all featured Release 5). The EPOC32 operating system, at the time simply referred to as EPOC, was later renamed Symbian OS. Adding to the confusion with names, before the change to Symbian, EPOC16 was often referred to as SIBO to distinguish it from the "new" EPOC. Despite the similarity of the names, EPOC32 and EPOC16 were completely different operating systems, EPOC32 being written in C++ from a new codebase with development beginning during the mid 1990s. EPOC32 was a pre-emptive multitasking, single user operating system with memory protection, which encourages the application developer to separate their program into an engine and an interface. The Psion line of PDAs come with a graphical user interface called EIKON which is specifically tailored for handheld machines with a keyboard (thus looking perhaps more similar to desktop GUIs than palmtop GUIs[82]). However, one of EPOC's characteristics is the ease with which new GUIs can be developed based on a core set of GUI classes, a feature which has been widely explored from Ericsson R380 and onwards.

	EPOC32 was originally developed for the ARM family of processors, including the ARM7, ARM9, StrongARM and Intel's XScale, but can be compiled towards target devices using several other processor types. During the development of EPOC32, Psion planned to license EPOC to third-party device manufacturers, and spin off its software division as Psion Software. One of the first licensees was the short-lived *Geofox*, which halted production with less than 1,000 units sold. Ericsson marketed a rebranded Psion Series 5mx called the *MC218*, and later created the EPOC Release 5.1 based smartphone, the *R380*. Oregon Scientific also released a budget EPOC device, the *Osaris* (notable as the only EPOC device to ship with Release 4). Work started on the 32-bit version in late 1994. The Series 5 device, released in June 1997, used the first iterations of the EPOC32 OS, codenamed "Protea", and the "Eikon" graphical user interface. The Oregon Scientific Osaris was the only PDA to use the ER4. The Psion Series 5mx, Psion Series 7, Psion Revo, Diamond Mako, Psion netBook and Ericsson MC218 were released in 1999 using ER5. A phone project was announced at CeBIT, the Phillips Illium/Accent, but did not achieve a commercial release. This release has been retrospectively dubbed Symbian OS 5. The first phone using ER5u, the Ericsson R380 was released in November 2000. It was not an 'open' phone – software could not be installed. Notably, a number of never-released Psion prototypes for next generation PDAs, including a Bluetooth Revo successor codenamed "Conan" were using ER5u. The 'u' in the name refers to the fact that it supported Unicode. In June 1998, Psion Software became Symbian Ltd., a major joint venture between Psion and phone manufacturers Ericsson, Motorola, and Nokia. As of Release 6, EPOC became known simply as Symbian OS.
Symbian OS 6.0 and 6.1	The OS was renamed Symbian OS and was envisioned as the base for a new range of smartphones. This release is sometimes called ER6. Psion gave 130 key staff to the new company and retained a 31% shareholding in the spin-off. The first 'open' Symbian OS phone, the Nokia 9210 Communicator, was released in June 2001. Bluetooth support was added. Almost 500,000 Symbian phones were shipped in 2001, rising to 2.1 million the following year. Development of different UIs was made generic with a "reference design strategy" for either 'smartphone' or 'communicator' devices, subdivided further into keyboard- or tablet-based designs. Two reference UIs (DFRDs or Device Family Reference Designs) were shipped – Quartz and Crystal. The former was merged with Ericsson's 'Ronneby' design and became the basis for the UIQ interface; the latter reached the market as the Nokia Series 80 UI. Later DFRDs were Sapphire, Ruby, and Emerald. Only Sapphire came to market, evolving into the Pearl DFRD and finally the Nokia Series 60 UI, a keypad-based 'square' UI for the first true smartphones. The first one of them was the Nokia 7650 smartphone (featuring Symbian OS 6.1), which was also the first with a built-in camera, with VGA (0.3 Mpx = 640×480) resolution. Other notable S60 Symbian 6.1 devices are the Nokia 3650, the short lived Sendo X and Siemens SX1 - the first and the last Symbian phone from Siemens. Despite these efforts to be generic, the UI was clearly split between competing companies: Crystal or Sapphire was Nokia, Quartz was Ericsson. DFRD was abandoned by Symbian in late 2002, as part of an active retreat from UI development in favour of 'headless' delivery. Pearl was given to Nokia, Quartz development was spun off as UIQ Technology AB, and work with Japanese firms was quickly folded into the MOAP standard.
Symbian OS 7.0 and 7.0s	First shipped in 2003. This is an important Symbian release which appeared with all contemporary user interfaces including UIQ (Sony Ericsson P800, P900, P910, Motorola A925, A1000), Series 80 (Nokia 9300, 9500), Series 90 (Nokia 7710), Series 60 (Nokia 3230, 6260, 6600, 6670, 7610) as well as several FOMA phones in Japan. It also added EDGE support and IPv6. Java support was changed from pJava and JavaPhone to one based on the Java ME standard. One million Symbian phones were shipped in Q1 2003, with the rate increasing to one million a month by the end of 2003. Symbian OS 7.0s was a version of 7.0 special adapted to have greater backward compatibility with Symbian OS 6.x, partly for compatibility between the Communicator 9500 and its predecessor the Communicator 9210. In 2004, Psion sold its stake in Symbian. The same year, the first worm for mobile phones using Symbian OS, *Cabir*, was developed, which used Bluetooth to spread itself to nearby phones. See Cabir and Symbian OS threats.
Symbian OS 8.0	First shipped in 2004, one of its advantages would have been a choice of two different kernels (EKA1 or EKA2). However, the EKA2 kernel version did not ship until Symbian OS 8.1b. The kernels behave more or less identically from user-side, but are internally very different. EKA1 was chosen by some manufacturers to maintain compatibility with old device drivers, while EKA2 was a real-time kernel. 8.0b was deproductised in 2003. Also included were new APIs to support CDMA, 3G, two-way data streaming, DVB-H, and OpenGL ES with vector graphics and direct screen access.

Symbian OS 8.1	An improved version of 8.0, this was available in 8.1a and 8.1b versions, with EKA1 and EKA2 kernels respectively. The 8.1b version, with EKA2's single-chip phone support but no additional security layer, was popular among Japanese phone companies desiring the real-time support but not allowing open application installation. The first and maybe the most famous smartphone featuring Symbian OS 8.1a was Nokia N90 in 2005, Nokia's first in Nseries.
Symbian OS 9.0	Symbian OS 9.0 was used for internal Symbian purposes only. It was de-productised in 2004. 9.0 marked the end of the road for EKA1. 8.1a is the final EKA1 version of Symbian OS. Symbian OS has generally maintained reasonable binary code compatibility. In theory the OS was BC from ER1-ER5, then from 6.0 to 8.1b. Substantial changes were needed for 9.0, related to tools and security, but this should be a one-off event. The move from requiring ARMv4 to requiring ARMv5 did not break backwards compatibility.
Symbian OS 9.1	Released early 2005. It includes many new security related features, including platform security module facilitating mandatory code signing. The new ARM EABI binary model means developers need to retool and the security changes mean they may have to recode. S60 platform 3rd Edition phones have Symbian OS 9.1. Sony Ericsson is shipping the M600 and P990 based on Symbian OS 9.1. The earlier versions had a defect where the phone hangs temporarily after the owner sent a large number of SMS'es. However, on 13 September 2006, Nokia released a small program to fix this defect.[83] Support for Bluetooth 2.0 was also added. Symbian 9.1 introduced capabilities and a Platform Security framework. To access certain APIs, developers have to sign their application with a digital signature. Basic capabilities are user-grantable and developers can self-sign them, while more advanced capabilities require certification and signing via the Symbian Signed [36] program, which uses independent 'test houses' and phone manufacturers for approval. For example, file writing is a user-grantable capability while access to Multimedia Device Drivers require phone manufacturer approval. A TC TrustCenter ACS Publisher ID certificate is required by the developer for signing applications.
Symbian OS 9.2	Released Q1 2006. Support for OMA Device Management 1.2 (was 1.1.2). Vietnamese language support. S60 3rd Edition Feature Pack 1 phones have Symbian OS 9.2. Nokia phones with Symbian OS 9.2 OS include the Nokia E71, Nokia E90, Nokia N95, Nokia N82, Nokia N81 and Nokia 5700.
Symbian OS 9.3	Released on 12 July 2006. Upgrades include improved memory management and native support for Wifi 802.11, HSDPA. The Nokia E72, Nokia 5730 XpressMusic, Nokia N79, Nokia N96, Nokia E52, Nokia E75, Nokia 5320 XpressMusic, Sony Ericsson P1 and others feature Symbian OS 9.3.
Symbian OS 9.4	Announced in March 2007. Provides the concept of demand paging which is available from v9.3 onwards. Applications should launch up to 75% faster. Additionally, SQL support is provided by SQLite. Ships with the Samsung i8910 Omnia HD, Nokia N97, Nokia N97 mini, Nokia 5800 XpressMusic, Nokia 5530 XpressMusic, Nokia 5228, Nokia 5230, Nokia 5233, Nokia 5235, Nokia C6-00, Nokia X6, Sony Ericsson Satio, Sony Ericsson Vivaz and Sony Ericsson Vivaz Pro, Micromax x265. Used as the basis for Symbian^1, the first Symbian platform release. The release is also better known as S60 5th edition, as it is the bundled interface for the OS.
Symbian^2	Symbian^2 is a version of Symbian that only used by Japanese manufacturers, started selling in Japan market since May of 2010.[84] The version is not used by Nokia.[85]
Symbian^3 (Symbian OS 9.5) and Symbian Anna	Symbian^3 is an improvement over previous S60 5th Edition and features single touch menus in the user interface, as well as new Symbian OS kernel with hardware-accelerated graphics; further improvements will come in the first half of 2011 including portrait qwerty keyboard, a new browser and split-screen text input. Nokia announced that updates to Symbian^3 interface will be delivered gradually, as they are available; Symbian^4, the previously planned major release, is now discontinued and some of its intended features will be incorporated into Symbian^3 in successive releases, starting with Symbian Anna.
Nokia Belle (Symbian OS 10.1)	In the summer of 2011 videos showing an early leaked version of Symbian Belle (original name of Nokia Belle) running on a Nokia N8 were published on YouTube[86]. On 24 August 2011, Nokia announced it officially for three new smartphones, the Nokia 600 (later replaced by Nokia 603), Nokia 700, and Nokia 701[87]. Nokia officially renamed Symbian Belle to Nokia Belle in a company blog post.[88][89] Nokia Belle adds to the Anna improvements with a pull-down status/notification bar, deeper near field communication integration, free-form re-sizable homescreen widgets, and six homescreens instead of the previous three. As of 7 February 2012, Nokia Belle update is available for most phone models through Nokia Suite, coming later to Australia. Users can check the availability on Nokia homepage[90]. On 1 March 2012, Nokia announced a Feature Pack 1 update for Nokia Belle which will be available as an update to Nokia 603, 700, 701 (excluding others), and for Nokia 808 PureView natively.[91]

References

[1] Nokia and Accenture Finalize Symbian Software Development and Support Services Outsourcing Agreement (http://newsroom.accenture.com/news/nokia-and-accenture-finalize-symbian-software-development-and-support-services-outsourcing.htm)
[2] Lextrait, Vincent (January 2010). "The Programming Languages Beacon, v10.0" (http://www.lextrait.com/Vincent/implementations.html). . Retrieved 5 January 2010.
[3] Not Open Source, just Open for Business (http://symbian.nokia.com/blog/2011/04/04/not-open-source-just-open-for-business/). symbian.nokia.com (2011-04-04). Retrieved on 2011-09-25.
[4] History of Symbian
[5] Lee Williams Symbian on Intel's Atom architecture (http://web.archive.org/web/20090419214755/http://blog.symbian.org/2009/04/16/symbian-on-intels-atom/). blog.symbian.org. 16 April 2009
[6] Uikon-Eikon-Avkon-Qikon - Nokia Developer Wiki (http://www.developer.nokia.com/Community/Wiki/Uikon-Eikon-Avkon-Qikon)
[7] http://worldwide.nokia.com/belle/
[8] Lunden, Ingrid (2011-09-30). "Symbian Now Officially No Longer Under The Wing Of Nokia, 2,300 Jobs Go" (http://moconews.net/article/419-symbian-now-officially-no-longer-under-the-wing-of-nokia-2300-jobs-go/). *moconews.net*. . Retrieved 30 September 2011.
[9] Nokia announces Symbian 'Anna' update for N8, E7, C7 and C6-01; first of a series of updates (video) (http://www.engadget.com/2011/04/12/nokia-announces-symbian-anna-update-for-n8-e7-c7-and-c6-01/). Engadget. Retrieved on 2011-09-25.
[10] Nokia announces Symbian Belle alongside three new devices (http://www.engadget.com/2011/08/24/nokia-announces-symbian-belle-running-on-three-new-devices/). Engadget. Retrieved on 2011-09-25.
[11] "infoSync Interviews Nokia Nseries Executive" (http://www.infosyncworld.com/news/n/11070.html). Infosyncworld.com. 2010-06-24. . Retrieved 2010-08-12.
[12] 100 Million Club H1 2010 (http://www.visionmobile.com/blog/2010/10/smart-feature-phones-the-unbalanced-equation-100-million-club-series/). VisionMobile (2010-10-18). Retrieved on 2011-09-25.
[13] RIP: Symbian (http://www.engadget.com/2011/02/11/rip-symbian/). Engadget. Retrieved on 2011-09-25.
[14] Epstein, Zach. (2011-06-23) Symbian is officially no longer Nokia's problem (http://www.bgr.com/2011/06/23/symbian-is-officially-no-longer-nokias-problem/). Bgr.com. Retrieved on 2011-09-25.
[15] Symbian OS – one of the most successful failures in tech history (http://eu.techcrunch.com/2010/11/08/guest-post-symbian-os-one-of-the-most-successful-failures-in-tech-history/). TechCrunch.com. 8 November 2010
[16] Symbian Foundation (2010-02-04), *Symbian Completes Biggest Open Source Migration Project Ever* (http://www.symbian.org/news-and-media/2010/02/04/symbian-completes-biggest-open-source-migration-project-ever), , retrieved 2010-02-07
[17] Menezes, Gary. (2010-09-11) Symbian OS, Now Fully Open Source (http://www.watblog.com/2010/02/06/symbian-os-now-fully-open-source). Watblog.com. Retrieved on 2011-09-25.
[18] "Symbian Foundation website, members section" (http://www.symbian.org/members). .
[19] "Developer Economics 2011" (http://www.visionmobile.com/blog/2011/06/developer-economics-2011-winners-and-losers-in-the-platform-race/). .
[20] symbian-dump | Download symbian-dump software for free at. Sourceforge.net. Retrieved on 2011-09-25.
[21] symbian-incubation-projects – Symbian Incubation Projects – Google Project Hosting (http://code.google.com/p/symbian-incubation-projects/). Code.google.com. Retrieved on 2011-09-25.
[22] Nokia PR (21 October 2010). "Nokia further refines development strategy to unify environments for Symbian and MeeGo" (http://www.nokia.com/press/press-releases/showpressrelease?newsid=1453894). . Retrieved 2010-11-05.
[23] AllAboutSymbian (26 October 2010). "The future of the Symbian platform" (http://www.allaboutsymbian.com/features/item/12223_The_future_of_the_Symbian_plat.php). . Retrieved 2010-11-05.
[24] Nokia PR (24 May 2006). "Nokia releases 'Web Browser for S60' engine code to open source community" (http://www.nokia.com/A4136002?newsid=1052589). *press.nokia.com*. . Retrieved 2007-03-21.
[25] Browser and Maps updates for many S60 3rd Edition and S60 5th Edition phones (http://www.allaboutsymbian.com/news/item/13056_Many_S60_3rd_Edition_and_S60_5.php). All About Symbian (2011-06-29). Retrieved on 2011-09-25.
[26] "Symbian — Qt – A cross-platform application and UI framework" (http://qt.nokia.com/products/platform/symbian/). Qt.nokia.com. . Retrieved 2010-08-12.
[27] Nokia Developer (2010-06-18), *Nokia Qt SDK* (http://developer.nokia.com/Develop/Qt/), , retrieved 2012-01-20
[28] Apps:Mobile Web Apps in a Nutshell (http://www.symlab.org/wiki/index.php/Apps:Mobile_Web_Apps_in_a_Nutshell). symlab.org wiki
[29] Nokia Developer – Web (http://www.forum.nokia.com/Technology_Topics/Web_Technologies/Web_Runtime/). Forum.nokia.com. Retrieved on 2011-09-25.
[30] "Qt Labs Blogs " Nokia Qt SDK 1.0 released" (http://labs.trolltech.com/blogs/2010/06/23/nokia-qt-sdk-10-released/). Labs.trolltech.com. . Retrieved 2010-08-12.
[31] "Qt Labs Blogs " Qt Simulator is going public" (http://labs.trolltech.com/blogs/2010/05/31/qt-simulator-is-going-public/). Labs.trolltech.com. . Retrieved 2010-08-12.
[32] "Symbian developer community" (http://developer.symbian.org). Developer.symbian.org. 2010-01-27. . Retrieved 2010-08-12.
[33] http://www.oracle.com/appforge/

[34] http://www.redfivelabs.com/content/WhyNet60.aspx
[35] Tom Sutcliffe and Jason Barrie Morley Xcode Symbian support (http://symbian-xcode-plugin.tigris.org/). Symbian-xcode-plugin.tigris.org. Retrieved on 2011-09-25.
[36] http://www.symbiansigned.com
[37] "Capabilities (Symbian Signed) – Symbian Developer Community" (http://developer.symbian.org/wiki/index.php/Capabilities_(Symbian_Signed)). Developer.symbian.org. . Retrieved 2010-08-12.
[38] Nokia Developer News | Nokia Now Signing Symbian Apps for Free – Nokia Developer Blogs (http://blogs.forum.nokia.com/blog/nokia-developer-news/2010/08/16/nokia-now-signing-symbian-apps-for-free). Blogs.forum.nokia.com (2010-08-16). Retrieved on 2011-09-25.
[39] "Symbian developer community – technology domains" (http://developer.symbian.org/main/source/technology_domains/index.php). Developer.symbian.org. . Retrieved 2010-08-12.
[40] "Symbian developer community – packages" (http://developer.symbian.org/main/source/packages/index.php). Developer.symbian.org. . Retrieved 2010-08-12.
[41] "Symbian System Model – Symbian Developer Community" (http://developer.symbian.org/wiki/index.php/Symbian_System_Model). Developer.symbian.org. . Retrieved 2010-08-12.
[42] Introducing EKA2, by Jane Sales with Martin Tasker (http://media.wiley.com/product_data/excerpt/47/04700252/0470025247.pdf). (PDF) . Retrieved on 2011-09-25.
[43] "Six Years of Symbian Produces 100 Models and 100 Million Shipments" (http://www.thesmartpda.com/50226711/six_years_of_symbian_produces_100_models_and_100_million_shipments.php). The Smart PDA. 2006-11-17. . Retrieved 2012-05-23.
[44] Symbian Foundation Adds New Member, Nuance (http://news.softpedia.com/news/Symbian-Foundation-Adds-New-Member-Nuance-117209.shtml). News.softpedia.com (2009 07 21). Retrieved on 2011 09 25.
[45] http://www.symbianblogs.com/history/110-symbian-2--history.html
[46] Nokia N8 User Agent Profile (http://nds.nokia.com/uaprof/NN8-00r100-3G.xml). Nds.nokia.com (1999-02-22). Retrieved on 2011-09-25.
[47] Nokia launches mobile TV | Nokia Conversations – The official Nokia Blog (http://conversations.nokia.com/2010/09/09/nokia-launches-mobile-tv/). Conversations.nokia.com (2010-09-09). Retrieved on 2011-09-25.
[48] "F-07B Instruction Manual '10.5" (http://www.nttdocomo.co.jp/english/binary/pdf/support/trouble/manual/download/f07b/F-07B_E_All.pdf) (PDF). *docomo STYLE series*. NTT DoCoMo. May 2010. . Retrieved 2012-05-23.
[49] "Any plans to have SilverLight for Symbian^3 (Nokia N8,E7,C7) ?" (http://forums.silverlight.net/t/214728.aspx/1?Any+plans+to+have+SilverLight+for+Symbian+3+Nokia+N8+E7+C7+). *Mobile / Silverlight for Nokia Symbian*. Silverlight.NET Forums. 2011-04-27. . Retrieved 2012-05-23.
[50] Psychlist1972 (2010-07-06). "Silverlight for Nokia Symbian RTW Now Available" (http://forums.silverlight.net/p/190204/438158.aspx/1?Silverlight+for+Nokia+Symbian+RTW+Now+Available). *Mobile / Silverlight for Nokia Symbian*. Silverlight.NET Forums. . Retrieved 2012-05-23.
[51] Obsolete (http://www.silverlight.net/getstarted/devices/symbian/). Silverlight.NET. Retrieved on 2011-09-25.
[52] http://forums.silverlight.net/t/190330.aspx/1?support+for+s60v3
[53] Inside Symbian SQL: A Mobile Developer's Guide to SQLite |

By Ivan Litovski, Richard Maynard, 2010, page 9

[54] (PDF) *SH-08C Instruction Manual '11.3* (http://www.nttdocomo.co.jp/english/binary/pdf/support/trouble/manual/download/sh08c/SH-08C_E_All.pdf), NTT DoCoMo, March 2011, , retrieved 2012-05-23
[55] "Help – Eclipse Platform" (http://library.forum.nokia.com/index.jsp?topic=/Web_Developers_Library/GUID-B7D6EFF3-16E6-45D5-9B74-8333BA83FC7B.html). library.forum.nokia.com. . Retrieved 2011-09-25.
[56] http://www.twitter.com/docomo_map_navi
[57] http://dmapnavi.jp/
[58] on your Mobile (http://www.skype.com/intl/en/get-skype/on-your-mobile/?cm_mmc=m102). Skype. Retrieved on 2011-09-25.
[59] Screenshot for Symbian OS | AntonyPranata.com 2.0 (http://www.antonypranata.com/screenshot/screenshot-symbian-os). Antonypranata.com. Retrieved on 2011-09-25.
[60] Horikawa, Kyoko (2010-06-01). "NTT DoCoMo releases S^2 devices" (http://web.archive.org/web/20100824165132/http://blog.symbian.org/2010/06/01/ntt-docomo-releases-s2-devices/). *Symbian Blog*. Symbian.org. Archived from the original (http://blog.symbian.org/2010/06/01/ntt-docomo-releases-s2-devices/) on 2010-08-24. .
[61] Asuk Ustundag, Sennur (2010-10-07). "SYMBIAN Devices, Hardware & Software Requirements, Basic App Development" (http://webhost.bridgew.edu/jsantore/Teaching-Archive/Fall2010/MobileDev/SYMBIAN_MDD_sennur.pdf) (PDF). Bridgewater State University. p. 6. . Retrieved 2012-05-23.
[62] "Symbian^2 platform used in eleven new models of NTT DoCoMo FOMA 3G handsets" (http://www.symbianone.com/content/view/7108/). SymbianOne. . Retrieved 2010-11-10.
[63] "Samsung OMNIAHD Dazzles at Mobile World Congress with Its HD Brilliance" (http://www.samsung.com/uk/news/newsPreviewRead.do?news_seq=12421). United Kingdom: Samsung.com. . Retrieved 2011-09-25.
[64] Pettey, Christy. "Gartner Says Worldwide Mobile Device Sales to End Users Reached 1.6 Billion Units in 2010; Smartphone Sales Grew 72 Percent in 2010" (http://www.gartner.com/it/page.jsp?id=1543014). Gartner.com. . Retrieved 2011-03-10.

[65] "Majority of smart phones now have touch screens (Canalys press release: r2010021)" (http://www.canalys.com/pr/2010/r2010021.html). Canalys.com. 2010-02-08. . Retrieved 2010-08-12.

[66] "BBC News – Google Android phone shipments increase by 886%" (http://www.bbc.co.uk/news/technology-10839034). Bbc.co.uk. 2010-08-02. . Retrieved 2010-08-12.

[67] Nokia Leading Smartphone Market with 56%, While Symbian's Share of OS Market Is Set to Fall | Press Release (http://www.abiresearch.com/press/826). ABI Research. Retrieved on 2011-09-25.

[68] Gartner Says Sales of Mobile Devices in Second Quarter of 2011 Grew 16.5 Percent Year-on-Year; Smartphone Sales Grew 74 Percent (http://www.gartner.com/it/page.jsp?id=1764714). Gartner.com. Retrieved on 2011-09-25.

[69] Nokia and Microsoft enter strategic alliance on Windows Phone, Bing, Xbox Live and more (http://www.engadget.com/2011/02/11/nokia-and-microsoft-enter-strategic-alliance-on-windows-phone-b/). Engadget. Retrieved on 2011-09-25.

[70] Woods, Ben. (2010-10-01) Samsung to drop Symbian support | Wireless – CNET News (http://news.cnet.com/8301-1035_3-20018315-94.html). News.cnet.com. Retrieved on 2011-09-25.

[71] Meyer, David. (2008-11-03) Motorola ditches Symbian, announces 3,000 layoffs | Networking | ZDNet UK (http://www.zdnet.co.uk/news/networking/2008/11/03/motorola-ditches-symbian-announces-3000-layoffs-39539063/). Zdnet.co.uk. Retrieved on 2011-09-25.

[72] Mello, John P.. (2010-10-15) Sony Ditches Symbian (http://www.pcworld.com/article/208000/sony_ditches_symbian.html#tk.mod_rel). PCWorld. Retrieved on 2011-09-25.

[73] Mobile-reviews. Review of Nokia E7. http://mobile-review.com/review/nokia-e7-en.shtml

[74] Mobile browser comparison, November 2010 (http://www.allaboutsymbian.com/features/item/12323_Mobile_browser_comparison_Nove.php). Allaboutsymbian.com (2010-11-25). Retrieved on 2011-09-25.

[75] Meyer, David (9 November 2010). "Nokia times first [http://www.symbians60v3.info/ Symbian (http://www.zdnet.co.uk/news/mobile-it/2010/11/09/nokia-times-first-symbian-updates-for-early-2011-40090806/) updates for 'early 2011'"]. ZDNet UK. . Retrieved 4 January 2011.

[76] http://www.symbians60v3.info/

[77] Hacker plants back door in Symbian firmware – The H Security: News and Features (http://www.h-online.com/security/news/item/Hacker-plants-back-door-in-Symbian-firmware-1149926.html). H-online.com (2010-12-08). Retrieved on 2011-09-25.

[78] Hacker Creates Modified Symbian S60 Firmware with Hidden Back Door (http://www.livehacking.com/2010/12/10/hacker-creates-modified-symbian-s60-firmware-with-hidden-back-door/). Live Hacking (2010-12-10). Retrieved on 2011-09-25.

[79] Nokia's S60 3rd Ed security has been hacked? (http://www.symbian-freak.com/news/008/03/s60_3rd_ed_has_been_hacked.htm), Symbian Freak

[80] "S60 v3 Hacking – Mission accomplished, FP1 hacked!" (http://www.symbian-freak.com/news/008/03/s60_3rd_ed_feature_pack_1_has_been_hacked.htm). Symbian Freak (2008-03-27). Retrieved on 2011-09-25.

[81] *Sibo3a screenshots* (http://www.guidebookgallery.org/screenshots/sibo3a), Guide Book Gallery, .

[82] Marcin Wichary. "GUIdebook - Screenshots - EPOC R5/Psion Revo" (http://www.guidebookgallery.org/screenshots/epocr5). Guidebookgallery.org. . Retrieved 2010-08-12.

[83] Solution to Nokia Slow SMS / Hang Problem (http://www.kejut.com/nokiasms)

[84] http://www.allaboutsymbian.com/news/item/11613_First_Symbian2_phones_ship_in_.php

[85] http://www.reghardware.com/2010/02/02/nokia_symbian_roadmap/

[86] http://www.engadget.com/2011/08/17/symbian-belle-download-leaked-to-n8-community-quickly-pulled-fr/

[87] http://www.unwiredview.com/2011/08/24/nokia-600-700-and-701-announced-all-running-symbian-belle-and-coming-before-the-end-of-september/

[88] http://conversations.nokia.com/2011/12/21/nokia-belle-coming-soon/

[89] http://news.cnet.com/8301-13506_3-57346089-17/so-long-symbian-belle-hello-nokia-belle/

[90] http://europe.nokia.com/find-products/nokia-belle-update/nokia-belle-update-availability

[91] http://conversations.nokia.com/2012/03/01/all-about-nokia-belle-feature-pack-1/

Bibliography

- Morris, Ben (22 June 2007). *The Symbian OS architecture sourcebook : design and evolution of a mobile phone OS* (http://eu.wiley.com/WileyCDA/WileyTitle/productCd-0470018461.html). John Wiley & Sons. p. 630. ISBN 0-470-01846-1.

External links

- Symbian foundation blog (which the homepage redirects to) (http://symbian.org/)
- Symbian (http://www.ohloh.net/p/symbian/analyses/latest) on Ohloh
- Symbian (http://www.dmoz.org/Computers/Mobile_Computing/Symbian/Symbian_OS/) at the Open Directory Project
- Qt developer website (http://developer.qt.nokia.com/)
- Symbian C++ developer website (http://www.developer.nokia.com/Develop/Other_Technologies/Symbian_C++/)

Broadcom

Broadcom Corporation

Type	Public
Traded as	NASDAQ: BRCM [1] NASDAQ-100 Component S&P 500 Component
Industry	Semiconductors Electronics
Founded	August 1991
Founder(s)	Henry Nicholas Henry Samueli
Headquarters	Irvine, California, U.S.
Key people	Scott A. McGregor (President & CEO) Henry Samueli (CTO)
Products	Integrated Circuits Cable Converter Boxes Gigabit Ethernet Wireless networks Cable modems Mobile communications Network Switches Digital Subscriber Line Server farms Processors Bluetooth VoIP Near Field Communication GPS Metropolitan Area Network
Revenue	▲ US$ 7.39 billion (2011)
Operating income	▲ US$ 953 million (2011)
Net income	▲ US$ 927 million (2011)
Employees	~11,000 (Q2 2012)
Website	Broadcom.com [2]

Broadcom Corporation is a fabless semiconductor company in the wireless and broadband communication business. The company is headquartered in Irvine, California, USA. Broadcom was founded by a professor-student pair Henry Samueli and Henry T. Nicholas III from the University of California, Los Angeles (UCLA) at Los Angeles, California in 1991. In 1995, the company moved from its Westwood, California, office to Irvine, California.[3] In 1998, Broadcom became a public company on the NASDAQ exchange (ticker symbol: BRCM) and now employs approximately 11,000 people worldwide in more than 15 countries.

Broadcom headquarters at UC Irvine's University Research Park

Broadcom is among Gartner's Top 10 Semiconductor Vendors by revenue.[4] In 2011, Broadcom's total revenue was $7.39 billion. In 2011, Broadcom was No. 343 on the Fortune 500, climbing 117 places from its 2010 ranking of No. 460.[5] Broadcom first landed on the Fortune 500 in 2009. The Broadcom logo is inspired from the Mathematical Sinc function .

Products

Broadcom's product line spans computer and telecommunication networking: the company has products for enterprise/metropolitan high-speed networks, as well as products for SOHO (small-office, home-office) networks. Products include transceiver and processor ICs for Ethernet and wireless LANs, cable modems, digital subscriber line (DSL), servers, home networking devices (router, switches, port-concentrators) and cellular phones (GSM/GPRS/EDGE/W-CDMA). It is also known for a series of high-speed encryption co-processors, offloading this processor-intensive work to a dedicated chip, thus greatly speeding up tasks that utilize encryption. This has many practical benefits for e-commerce, and PGP or GPG secure communications.

The company also produces ICs for carrier access equipment, audio/video processors for digital set-top boxes and digital video recorders, Bluetooth and Wi-Fi transceivers, and RF receivers/tuners for satellite TV. Major customers include Apple Computer, Hewlett-Packard, Motorola, IBM, Dell, Lenovo, Linksys, Logitech, Nintendo, Nokia Siemens Networks, Nortel(Avaya), TiVo and Cisco Systems. In September 2011, Broadcom shut down its digital TV operations.[6] Broadcom also shut down its Blu-ray chip business. The closure of these businesses began on September 19, 2011.

NIC's and networking

All major hardware vendors include (amongst other vendors like Intel) Broadcom NIC's in their workstation and server-products. Even when hardware vendors implement ethernet NIC's on the motherboard they are marketed as Broadcom. For example the Dell blade-switch M610 has two embedded Gigabit NetXtreme 5709 NIC's[7]

Trident+ ASIC

Another large market is hardware for switches: some vendors offer switching equipment based on Broadcom hardware and firmware (e.g. Dell PowerConnect classics) while other well-known vendors do use the Broadcom hardware but write their own firmware. The latest Broadcom Trident+ ASIC is used in many high-speed/ 10Gb + switches from the largest switch-vendors such as: Cisco Nexus switches running NX-OS,[8]or the Dell Force10 running FTOS.[9][10] Also the Arista 7050S[11] the IBM/BNT 8264 and Juniper QFX3500[12] are all based on the Trident+ ASICs.

Consumer design components

Broadcom also provides components for a number of high-profile consumer devices:

- Broadcom supplies the WiFi+Bluetooth combo chip for Apple iPhone 3GS and iPod touch second generation.
- In Q2 2005, Broadcom Corporation announced it would be providing Nintendo its "online solution on a chip" as deployed in millions of notebooks and PDAs across the globe, enabling Nintendo 802.11b connectivity with DS and 802.11g for the Wii. More specifically, Broadcom would provide Bluetooth connectivity for Wii's controller.

Notable employees

- Henry Samueli
- Gottfried Ungerboeck, inventor of trellis coded modulation
- Henry T. Nicholas III
- Sophie Wilson, designer of the ARM CPU instruction set

Broadcom and Linux

Some open source drivers are available and included in the Linux kernel source tree for the 802.11b/g/a/n family of wireless chips Broadcom produces.[13] Since the release of the 2.6.26 kernel some Broadcom chips have kernel support but require external firmware to be built.

In 2003, the Free Software Foundation accused Broadcom of not complying with the GNU General Public License as Broadcom distributed GPL code in a driver for its 802.11g router chipset without making that code public. The chipset was adopted by Linksys which was later purchased by Cisco. Cisco eventually published source code for the firmware for its WRT54G wireless broadband router.[14][15]

In 2012, the Linux Foundation listed Broadcom as one of the Top 10 companies contributing to the development of the Linux Kernel for 2011, placing it in the top 5 percent of an estimated 226 contributing companies. The foundation's Linux Kernel Development report [16] also noted that, during the course of the year, Broadcom submitted 2,916 changes to the kernel.

Manufacturing

Broadcom is known as a *fabless* company. It outsources all semiconductor manufacturing to Asian merchant foundries, such as GlobalFoundries, Semiconductor Manufacturing International Corporation, Silterra, TSMC, and United Microelectronics Corporation. The company is based in Irvine, California in the University Research Park on the University of California, Irvine campus, after a 2007 move from its previous campus near the Irvine Spectrum. It has many other research and development sites including Silicon Fen, Cambridge (UK), Bangalore and Hyderabad in India, Richmond (near Vancouver) and Markham (near Toronto) in Canada, and Sophia Antipolis in France.

Stock options scandal

On July 14, 2006, Broadcom announced it had to subtract $750,000,000 from earnings due to stock options irregularities. On September 8, 2006 the amount was doubled to $1.5 billion. The company may also owe additional taxes.[17] On January 24, 2007, it announced a restatement of its financial results from 1998 to 2005 that totaled $2.22 billion.[18]

On May 15, 2008, Samueli, Broadcom CTO, resigned as chairman of the board and took of a leave of absence as Chief Technology Officer after being named in a civil complaint by the U.S. Securities and Exchange Commission (SEC).

On June 5, 2008, Broadcom co-founder and former CEO Henry Nicholas and former CFO William Ruehle were indicted on charges of illegal stock-option backdating. Nicholas was also indicted for violations of federal narcotics laws.[19] However, in December 2009, federal judge Cormac J. Carney threw out the options backdating charges against Nicholas and Ruehle after finding that federal prosecutors improperly tried to prevent three defense witnesses from testifying.[20]

Qualcomm Litigation and Settlement

On April 26, 2009, Broadcom settled four years of legal battles over wireless and other patents with Qualcomm Inc., another fabless semiconductor company headquartered in San Diego, California, USA.[21]

The deal ended the patent litigation as well as complaints of anti-competitive behavior before trade commissions in the U.S., Europe and South Korea. As part of the settlement, Qualcomm is paying $891 million in cash to Broadcom through April 2013.

In June 2007, the U.S. International Trade Commission blocked the import of new cell phone models based on particular Qualcomm microchips. They found that these Qualcomm microchips infringe patents owned by Broadcom.

BroadVoice

Broadcom authored its own VoIP codecs in 2002, and released them as open source with LGPL license in 2009:[22]

- **BroadVoice 16** with declared bitrate 16 kbit/s and audio sampling frequency 8 kHz
- **BroadVoice 32** with declared bitrate 32 kbit/s and sampling rate of 16 kHz (note however that X-Lite SIP phone's menu declares bitrate 80000 bit/s)

Acquisitions

In September 2011, Broadcom bought NetLogic Microsystems for a $3.7 billion deal in cash, excluding around $450 million of NetLogic employee shareholdings, which will transfer to Broadcom.[23]

Besides a big deal above, through the years, Broadcom has acquired many smaller companies to quickly enter new markets.[24]

Date	Acquired company	Amount	Expertise
January 1999	Maverick Networks	$104M in Stock	Multi-layer switches for corporate networks
April 1999	Epigram	$316M in stock	Home networking using telephone wiring
June 1999	Armedia Inc.	$67.2M in stock	Digital Video Decoders[25]
August 1999	HotHaus Technologies	$280M in stock	DSP software for VOIP
August 1999	Altocom	$180M in stock	Software modem software
January 2000	BlueSteel Networks	$123M in stock	Security processors
March 2000	Digital Furnace Corp	$136M in stock	Data compression software
March 2000	Stellar Semiconductor	$162M in stock	3D graphics processors
June 2000	Pivotal Technologies	$242M in stock	Digital video chips
July 2000	Innovent Systems	$500M in stock	Bluetooth radios
August 2000	Puyallup Integrated Circuit Company		IC design and IC macro blocks
July 2000	Altima Communications	$533M in stock	Networking chips
October 2000	Newport Communications	$1240M in stock	10Gbit Ethernet transceivers
October 2000	Silicon Spice	$1000M in stock	DSP chips for VOIP

November 2000	Element 14	$594M in stock	DSL chipsets
December 2000	Allayer Communications	$271M in stock	Enterprise and optical networking chips
December 2000	Sibyte	$2000M in stock	Broadband microprocessors
January 2001	VisionTech, Ltd.	$777M in stock	MPEG-2 compression/decompression of PVRs
January 2001	ServerWorks Corp.	$1003M in stock	I/O controllers for servers and workstations
July 2001	PortaTec Corporation		Mobile devices
July 2001	Kimalink		Wireless and mobile ICs
May 2002	Mobilink Telecom, Inc.	$5.6M shares of stock	Baseband processors for cellphones
March 2003	Gadzoox Networks	$5.8M in cash	Storage-area networks
January 2004	RAIDCore, Inc.	$16.5M in cash	RAID software
April 2004	M-Stream Inc.	$8.7M in cash and 27000 shares of stock	Technology to improve wireless reception
April 2004	Sand Video, Inc.	$77.5M in stock and $7.4M in cash	Video compression technology
April 2004	WIDCOMM, Inc.	$49M in cash	Software for Bluetooth systems
April 2004	Zyray Wireless, Inc.	$96M in stock	Baseband processors for WCDMA
September 2004	Alphamosaic, Ltd.	$123M in stock	Video processors for mobile devices
February 2005	Alliant Networks, Inc.		Cellular gateway products
March 2005	Zeevo, Inc.	$26.4M in cash and $2.6M in stock	Bluetooth headset products
July 2005	Siliquent Technologies, Inc.	$76M in cash	10Gbit Ethernet interface controllers
October 2005	Athena Semiconductors, Inc.	$21.6M in cash	Digital TV tuners and Wifi technology
January 2006	Sandburst Corporation	$75M in cash and $5M in stock	SOC chips for Ethernet packet switching
November 2006	LVL7 Systems, Inc.	$62M in cash	Networking software
May 2007	Octalica, Inc.	$31M in cash	Multimedia Over Coax technology
June 2007	Global Locate, Inc.	$146M in cash	GPS chips and software
March 2008	Sunext Design, Inc.	$48M in cash	Optical disk drive technologies
August 2008	AMD (DTV Processor Division)	$141.5M in cash (Original deal was $192.8M)[26]	Xilleon DTV processor chips, software and TV tuners
December 2009	Dune Networks[27]	$178M in cash	High speed network switches
February 2010	Teknovus[28]	$123M in cash	Ethernet Passive Optical Network (EPON) chipsets and software
June 2010	Innovision Research & Technology plc[29]	$47.5M in cash	Near field communication expertise and IP
October 2010	Beceem Communications[30]	$316M in cash	4G LTE/WiMax expertise
November 2010	Gigle Networks[31]	$75M in cash	Multimedia home networking

April 2011	Provigent Ltd.[32]	$313M in cash	Microwave Backhaul
May 2011	SC Square Ltd.[33]	$41.9M in cash	Israel-based security software developer
September 2011	NetLogic Microsystems	$3.7 billion	Next-generation Internet networks
March 2012	BroadLight[34]	$230M in cash	Israel-based fiber access PON developer

References

[1] http://www.nasdaq.com/symbol/brcm

[2] http://www.broadcom.com/

[3] Kotkin, Joel (January 24, 1999). "Grass Roots Business; A Place To Please The Techies - New York Times" (http://query.nytimes.com/gst/fullpage.html?res=9A07E1DB1F30F937A15752C0A96F958260&scp=23&sq=henry+nicholas+broadcom&st=nyt). The New York Times. . Retrieved 2008-04-25.

[4] Deffree, Suzanne (April 19, 2011). "Broadcom moves on to top 10 list as 2010 semi revenue records more than 30% growth" (http://www.edn.com/article/517893-Broadcom_moves_on_to_top_10_list_as_2010_semi_revenue_records_more_than_30_growth.php). EDN.com. . Retrieved 2011-05-05.

[5] "Fortune 500: Our Annual Ranking of America's Largest Corporations" (http://money.cnn.com/magazines/fortune/fortune500/2011/snapshots/11071.html). *CNN*. May 9, 2011. . Retrieved 2011-05-12.

[6] Junko Yoshida, EE Times. " Broadcom closes DTV, Blu-ray chip businesses (http://www.edn.com/article/519426-Broadcom_closes_DTV_Blu_ray_chip_businesses.php)." September 22, 2011. Retrieved October 4, 2011.

[7] Technical specs on the Dell PowerEdge M610 (http://www.dell.com/us/business/p/poweredge-m610/pd?~ck=), visited 27 Januari, 2012

[8] Cisco rolls out Nexus 3000 (http://www.computerworlduk.com/news/networking/3263636/cisco-rolls-out-nexus-3000-switch-for-low-latency-stock-trading-traffic/), visited 28 January 2012

[9] The Register on Force10 cranks Ethernet switches to 40 Gigabits (http://www.theregister.co.uk/2011/04/26/force10_networks_switch_upgrade/), 23 April 2011; visited 28 January 2012

[10] Jason Edelman Blog on 40 Gbps datacenter switching (http://www.jedelman.com/1/post/2011/12/40gbe-data-center-switching.html), 10 December, 2011. Retrieved 17 May, 2012

[11] The Register: Arista punts 10/40 GbE juice-sipper (http://www.theregister.co.uk/2011/03/28/arista_7050s_7124sx_switches/), visited 18 May, 2012

[12] Lightreading Re:Some Pizza (http://www.lightreading.com/messages.asp?piddl_msgthreadid=239626&piddl_msgorder=asc&piddl_msgid=347063#msg_347063), 30 April, 2012. Visited: 18 May, 2012

[13] b43 Sipsolutions.net (http://linuxwireless.sipsolutions.net/en/users/Drivers/b43), Linux Wireless

[14] "Linksys routers caught up in open source dispute" (http://searchnetworking.techtarget.com/news/article/0,289142,sid7_gci932666,00.html). *TechTarget*. 2003-10-20. .

[15] "Linux's Hit Men - Forbes.com" (http://www.forbes.com/2003/10/14/cz_dl_1014linksys.html). *Forbes*. . Retrieved 28 January 2012.

[16] http://go.linuxfoundation.org/who-writes-linux-2012

[17] "Broadcom's Options Bombshell" (http://www.businessweek.com/investor/content/sep2006/pi20060908_687749.htm?campaign_id=rss_null). *BusinessWeek*. 2006-09-09. . Retrieved 2006-09-09.

[18] "A $2.2 Billion Charge at Broadcom" (http://www.nytimes.com/2007/01/24/technology/24broadcom.html). 2007-01-24. . Retrieved 2012-02-15.

[19] "Drugs, hookers and cranked customers: Ex-Broadcom boss indicted" (http://www.theregister.co.uk/2008/06/05/henry_nicholas_indicted/). *The Register*. 2008-06-05. . Retrieved 2008-06-06.

[20] Flaccus, Gillian. Broadcom backdating case dismissed (http://articles.sfgate.com/2009-12-16/business/17224232_1_henry-nicholas-iii-broadcom-co-founder-henry-samueli-fair-trial). Associated Press via San Francisco Chronicle, 2009-12-16.

[21] Jones, Ashby (April 27, 2009). "All Quiet on the Western Front: Broadcom, Qualcomm Reach $891M Deal" (http://blogs.wsj.com/law/2009/04/27/all-quiet-on-the-western-front-broadcom-qualcomm-reach-891m-deal/). http://blogs.wsj.com. . Retrieved 2011-08-06.

[22] "Broadcom offers LGPL Voice Codecs" (http://www.h-online.com/open/news/item/Broadcom-offers-LGPL-Voice-Codecs-855379.html). .

[23] "Broadcom buys NetLogic for $3.7bn" (http://www.ft.com/intl/cms/s/2/353278f2-dd40-11e0-b4f2-00144feabdc0.html?ftcamp=rss#axzz1XkzEPOpJ). . Retrieved September 12, 2011.

[24] A list of acquisitions (http://www.broadcom.com/company/strategicacquisitions.php)

[25] Broadcom Acquires Armedia, Maker of Digital Video Decoders | http://articles.latimes.com/1999/jun/02/business/fi-43234

[26] Broadcom Completes Acquisition of Digital TV Business from AMD for $50M less (http://www.broadcom.com/press/release.php?id=1218310&industry_id=4)

[27] Broadcom to buy Dune Networks for cloud switches (http://news.techworld.com/virtualisation/3207887/broadcom-to-buy-dune-networks-for-cloud-switches/)
[28] Broadcom to acquire Teknovus (http://www.broadcom.com/press/release.php?id=s449949/)
[29] Broadcom to enter NFC market, buys Innovision for $47.5m (http://www.nearfieldcommunicationsworld.com/2010/06/18/33993/broadcom-to-enter-nfc-market-buys-innovision-for-47-5m/)
[30] Broadcom.com (http://www.broadcom.com/press/release.php?id=s517947)
[31] Broadcom.com (http://www.broadcom.com/press/release.php?id=s532148)
[32] Broadcom.com (http://www.broadcom.com/press/release.php?id=s571519)
[33] / Broadcom Completes Acquisition of SC Square Ltd. (http://www.sunherald.com/2011/05/23/3135828/broadcom-completes-acquisition.html)
[34] / Broadcom Enters Agreement to Acquire BroadLight. (http://www.broadcom.com/press/release.php?id=s658602)

External links

- Broadcom homepage (http://www.broadcom.com/)
- 20th Century History of Broadcom corporation (http://www.referenceforbusiness.com/history2/75/Broadcom-Corporation.html)
- Broadcom Analysis on Wikinvest

USB_On-The-Go

USB On-The-Go, often abbreviated **USB OTG**, is a specification that allows USB devices such as digital audio players or mobile phones to act as a host allowing a USB flash drive, mouse, or keyboard to be attached and also connecting USB peripherals directly for communication purposes among them.

Architecture

Standard USB uses a master/slave architecture; a USB host acts as the protocol master, and a USB *device* acts as the slave. Only the host can schedule the configuration and data transfers over the link. The devices cannot initiate data transfers, they only respond to requests given by a host. OTG introduces the concept that a device can perform both the master and slave roles, and so subtly changes the terminology. With OTG, a *device* can be either a *host* (acting as the link master) or a *peripheral* (acting as the link slave). The device connected to the "A" end of the cable at start-up (known as the A-device) acts as the default host, while the "B" end acts as the default peripheral (known as the B-device).

USB On-The-Go does not preclude using a USB hub, but it describes host/peripheral role swapping only for the case of a one-to-one connection where two OTG devices are directly connected. Role swapping does not work through a standard hub, as one device will act as the host and the other as the peripheral until they are disconnected.

Specifications

USB OTG is part of a supplement[1] to the Universal Serial Bus (USB) 2.0 specification originally agreed upon in late 2001 and later revised.[2] The latest version of this supplement also defines behavior for an Embedded Host which has targeted abilities and the same USB Standard-A port used by PCs.

SuperSpeed OTG devices, Embedded Hosts and peripherals are supported through the USB On-The-Go and Embedded Host Supplement[3] to the USB 3.0 specification.

Protocols

The USB On-The-Go and Embedded Host Supplement to the USB 2.0 specification introduced three new protocols, Attach Detection Protocol (ADP), Session Request Protocol (SRP) and Host Negotiation Protocol (HNP).

- ADP allows an OTG device, Embedded host or USB device to determine attachment status in the absence of power on the USB bus. This enables both insertion based behavior and the possibility for a device to display attachment status. It does this by periodically measuring the capacitance on the USB port to determine whether there is another device attached, a dangling cable or no cable. When a change in capacitance, large enough to indicate device attachment is detected then an A-device will provide power to the USB bus and look for device connection. A B-device will generate SRP and wait for the USB bus to become powered.
- SRP allows both communicating devices to control when the link's power session is active; in standard USB, only the host is capable of doing so. That allows fine control over the power consumption, which is very important for battery operated devices such as cameras and mobile phones. The OTG or Embedded host can leave the USB link unpowered until the peripheral (which can be an OTG or standard USB device) asks it to start delivering power. OTG and Embedded hosts may not have much power to spare from their batteries, and leaving the USB link unpowered helps stretch battery life.
- HNP allows the two devices to exchange their host/peripheral roles, provided both are OTG dual-role devices. By using HNP for reversing host/peripheral roles, the USB OTG device is capable of acquiring control of data-transfer scheduling. Thus, any OTG device is capable of initiating data-transfer over USB OTG bus. The latest version of the supplement also introduced the idea of HNP polling whereby the device in host role periodically polls the peripheral, during an active session, to determine whether it wishes to become a host.

The main purpose of HNP is to accommodate users who have connected the A and B devices (see below) in the wrong direction for the task they want to perform. For example, a printer is connected as the A-device (host), but cannot function as a host for a particular camera, since it doesn't understand the camera's representation of print jobs. When that camera knows how to talk to the printer, the printer will use HNP to switch to the slave role, making the camera the host to the printer so that the user's pictures will get printed without juggling cables. These new OTG protocols cannot pass through a standard USB hub since they are based on physical electrical-signaling.

The USB On-The-Go and Embedded Host Supplement to the USB 3.0 specification introduces an additional protocol, Role Swap Protocol (RSP). This achieves the same purpose as HNP (i.e. role swapping) by extending standard mechanisms provided by the USB 3.0 specification. Products following the USB On-The-Go and Embedded Host Supplement to the USB 3.0 specification are also required to follow the USB 2.0 supplement in order to maintain backwards compatibility. SuperSpeed OTG devices (SS-OTG) are required to support RSP. SuperSpeed Peripheral Capable OTG devices (SSPC-OTG) are not required to support RSP since they can only operate at SuperSpeed as a peripheral; they have no SuperSpeed host and so can only role swap using HNP at USB 2.0 data rates.

Device roles

USB OTG defines two roles of devices: OTG A-device and OTG B-device. This terminology defines which side supplies power to the link, and which is initially the host. The OTG A-device is a power supplier, and an OTG B-device is a power consumer. The default link configuration is that A-device act as USB Host and B-device is a USB peripheral. The host and peripheral modes may be exchanged later by using HNP. Because every OTG controller supports both roles, they are often called "Dual-Role" controllers rather than "OTG controllers".

For integrated circuit (IC) designers, an attraction of USB OTG is the ability to get more USB capabilities with fewer gates. A "traditional" approach includes four controllers:

- USB high speed host controller based on EHCI (a register interface)
- Full/low speed host controller based on OHCI (another register interface)

- USB device controller, supporting both high and full speeds
- Fourth controller to switch the OTG root port between host and device controllers.

This means many gates to test and debug. Also, most gadgets must be a host only, or a device only. OTG hardware design merges all of these controllers into one dual-role controller, that is somewhat more complex than the device controller alone.

Targeted peripheral list

The targeted peripheral list or TPL applies to all targeted hosts which includes both OTG devices acting in a host role and embedded hosts. The aim of the TPL is for a manufacturer to list products supported by the targeted host in order to define what it needs to support in terms of output power, speeds, protocols and device classes. The TPL is intended such that hosts can be "targeted" at a particular product or application rather than being forced to be general purpose hosts like a PC.

Plug

OTG mini plugs

The original USB On-The-Go standard introduced a plug receptacle called mini-AB which was replaced by the micro-AB in later revisions (Revision 1.4 onwards). It could accept either a mini-A plug or a mini-B plug. Mini-A Adapters allowed connection to standard-A USB cables, coming from peripherals. The standard OTG cable had a mini-A plug on one side and a mini-B plug on the other (it could not have two plugs of the same type). The device that had a mini-A plugged in became an OTG A-device, and the one that had mini-B plugged became a B-device (see above). The type of the plug inserted was detected by the state of the pin ID (the mini-A plug has the ID pin *grounded* while the ID in the mini-B plug was *floating*). (There were also pure Mini-A plugs, used where a compact host port is needed but OTG was not supported.)

OTG Micro Plugs

With the introduction of the USB Micro Plug, a new plug receptacle called Micro-AB was also introduced. It can accept either a Micro-A plug or a Micro-B plug. Micro-A Adapters allow for connection to Standard-A plug type USB cables, as used on standard USB 2.0 Devices. An OTG product must have a single Micro-AB receptacle and no other USB receptacles.[4][5]

The OTG cable has a micro-A plug on one side, and a micro-B plug on the other (it cannot have two plugs of the same type). OTG adds a fifth pin to the standard USB connector, called the ID-pin; the micro-A plug has the ID pin *grounded*, while the ID in the micro-B plug is *floating*. The device that has a micro-A plugged in becomes an OTG A-device, and the one that has micro-B plugged becomes a B-device. The type of the plug inserted is detected by the state of the pin ID .

Three additional ID pin states are defined[4] at the nominal resistance values of 124 kΩ, 68 kΩ, and 36.5 kΩ, with respect to the ground pin. These permit the device to work with a USB Accessory Charger Adapter which allows the OTG device to be attached to both a charger and another device simultaneously.[6] These three states are used in the cases of:

- A charger and either no device or an A-device that is not asserting VBUS (not providing power) are attached. The OTG device is allowed to charge and initiate SRP but not connect.[6]
- A charger and an A-device that is asserting VBUS (is providing power) are attached. The OTG device is allowed to charge and connect but not initiate SRP.[6]
- A charger and a B-device are attached. The OTG device is allowed to charge and enter host mode.[6]

USB 3.0 introduced a backwards compatible, SuperSpeed extension of the Micro-AB receptacle and Micro-A and Micro-B plugs. These contain all of the pins in the USB 2.0 Micro and use the ID pin to identify the A-device and B-device roles. Additionally they contain the additional SuperSpeed pins.

OTG micro cables

When attached to a PC an OTG device requires a cable which has a USB Standard-A plug on one end and a Micro-B plug on the other end. In order to attach a peripheral to an OTG device the peripheral either needs to have a cable ending in a Micro-A plug which is inserted into the OTG device's Micro-AB receptacle or the OTG device itself needs an adapter cable which has a Micro-A plug on one end and a Standard-A receptacle on the other. The adapter cable enables any standard USB peripheral to be attached to an OTG device. In order to attach two OTG devices together requires either a cable with a Micro-B plug at one end and a Micro-A plug at the other or can be achieved using a combination of the PC cable and adapter cable.

Backward compatibility

USB OTG devices are backward-compatible with USB 2.0 (USB 3.0 for SuperSpeed OTG devices) and will behave as standard USB hosts or devices when connected to standard (non-OTG) USB devices. The main exception is that OTG hosts are only required to provide enough power for the products listed on the TPL, which may or may not be enough to connect to a peripheral which is not listed. A powered USB hub may sidestep the issue if supported since this will then provide its own power according to either the USB 2.0 or USB 3.0 specifications.

Some incompatibilities in both HNP and SRP were introduced between the 1.3 and 2.0 versions of the On-The-Go supplement which may lead to interoperability issues when using these protocols.

Charger compatibility

Some devices can charge their battery via their USB port; some can even detect a dedicated charger and draw more than 500 mA, which allows them to charge faster. OTG devices are not excluded from either of these options.[6]

References

[1] On-The-Go and Embedded Host Supplement to the USB 2.0 Specification (http://www.usb.org/developers/onthego) Revision 2.0 plus ECN and errata, July 14, 2011

[2] USB-On-the-Go-Specification Settled (http://www.heise.de/english/USB-On-the-Go-Specification-Settled--/newsticker/news/23699). *Heise.de*, Heinz Heise.

[3] On-The-Go and Embedded Host Supplement to the USB 3.0 Specification (http://www.usb.org/developers/onthego) Revision 1.0, July 1, 2011

[4] "Universal Serial Bus Revision 2.0 specification" (http://www.usb.org/developers/docs/usb_20_071411.zip). *On-The-Go and Embedded Host Supplement to the USB Revision 2.0 Specification plus ECN and errata*. USB Implementers Forum, Inc. 14 July 2011. . Retrieved 28 December 2010.

[5] "Universal Serial Bus Revision 2.0 specification" (http://www.usb.org/developers/docs/usb_20_071411.zip). *Universal Serial Bus Micro-USB Cables and Connectors Specification*. USB Implementers Forum, Inc. 4 April 2009. . Retrieved 28 December 2010.

[6] "Battery Charging Specification" (http://www.usb.org/developers/devclass_docs/batt_charging_1_1.zip). USB Implementers Forum, Inc. 15 April 2009. . Retrieved 23 September 2009.

External links

- Official website (http://www.usb.org/developers/onthego)
- USB On-The-Go Basics (http://www.maxim-ic.com/app-notes/index.mvp/id/1822)
- USB On-The-Go introduction (http://phonetopchart.com/2011/12/about-the-usb-on-the-go-technology/)

Bluetooth_profile

A **Bluetooth profile** is a specification regarding an aspect of Bluetooth-based wireless communication between devices. In order to use Bluetooth technology, a device must be compatible with the subset of Bluetooth profiles necessary to use the desired services. A Bluetooth profile resides on top of the Bluetooth Core Specification and (optionally) additional protocols. While the profile may use certain features of the core specification, specific versions of profiles are rarely tied to specific versions of the core specification. For example, there are HFP 1.5 implementations using both Bluetooth 2.0 and Bluetooth 1.2 core specifications.

The way a device uses Bluetooth technology depends on its profile capabilities. The profiles provide standards which manufacturers follow to allow devices to use Bluetooth in the intended manner. For the Bluetooth low energy stack according to Bluetooth V4.0 a special set of profiles applies.

At a maximum, each profile specification contains information on the following topics:

- Dependencies on other formats
- Suggested user interface formats
- Specific parts of the Bluetooth protocol stack used by the profile. To perform its task, each profile uses particular options and parameters at each layer of the stack. This may include an outline of the required service record, if appropriate.

This article summarizes the current definitions and possible applications of each profile.

List of profiles

The following profiles are defined and adopted by the Bluetooth SIG:

Advanced Audio Distribution Profile (A2DP)

This profile defines how high quality audio (stereo or mono) can be streamed from one device to another over a Bluetooth connection. For example, music can be streamed from a mobile phone, to a wireless headset, hearing aid & cochlear implant streamer, car audio, or from a laptop/desktop to a wireless headset.

A2DP was initially used in conjunction with an intermediate Bluetooth transceiver that connects to a standard audio output jack, encodes the incoming audio to a Bluetooth-friendly format, and sends the signal wirelessly to Bluetooth headphones that decode and play the audio. Bluetooth headphones, especially the more advanced models, often come with a microphone and support for the Headset (HSP), Hands-Free (HFP) and Audio/Video Remote Control (AVRCP) profiles.

A2DP is designed to transfer a uni-directional 2-channel stereo audio stream, like music from an MP3 player, to a headset or car radio.[1] This profile relies on AVDTP and GAVDP. It includes mandatory support for the low-complexity SBC codec (not to be confused with Bluetooth's voice-signal codecs such as CVSDM), and supports optionally: MPEG-1, MPEG-2, MPEG-4, AAC, and ATRAC, and is extensible to support manufacturer-defined codecs, such as apt-X. Some Bluetooth stacks enforce the SCMS-T digital rights management (DRM) scheme. In these cases, it is impossible to connect certain A2DP headphones for high quality audio.

Attribute Profile (ATT)

The ATT is a wire application protocol for Bluetooth Low Energy specification. It is closely related to Generic Attribute Profile (GATT).

Audio/Video Remote Control Profile (AVRCP)

This profile is designed to provide a standard interface to control TVs, Hi-fi equipment, etc. to allow a single remote control (or other device) to control all of the A/V equipment to which a user has access. It may be used in concert with A2DP or VDP.

It has the possibility for vendor-dependent extensions.

AVRCP has several versions with significantly increasing functionality:

- 1.0—Basic remote control commands (play/pause/stop, etc.)
- 1.3—all of 1.0 plus metadata and media-player state support
  - The status of the music source (playing, stopped, etc.)
  - Metadata information on the track itself (artist, track name, etc.).
- 1.4—all of 1.0, 1.3, plus media browsing capabilities for multiple media players
  - Browsing and manipulation of multiple players
  - Browsing of media metadata per media player, including a "Now Playing" list
  - Basic search capabilities

Basic Imaging Profile (BIP)

This profile is designed for sending images between devices and includes the ability to resize, and convert images to make them suitable for the receiving device. It may be broken down into smaller pieces:

Image Push

  Allows the sending of images from a device the user controls.

Image Pull

  Allows the browsing and retrieval of images from a remote device.

Advanced Image Printing

  print images with advanced options using the DPOF format developed by Canon, Kodak, Fujifilm, and Matsushita

Automatic Archive

  Allows the automatic backup of all the new images from a target device. For example, a laptop could download all of the new pictures from a camera whenever it is within range.

Remote Camera

  Allows the initiator to remotely use a digital camera. For example, a user could place a camera on a tripod for a group photo, use their phone handset to check that everyone is in frame, and activate the shutter with the user in the photo.

Remote Display

  Allows the initiator to push images to be displayed on another device. For example, a user could give a presentation by sending the slides to a video projector.

Basic Printing Profile (BPP)

This allows devices to send text, e-mails, vCards, or other items to printers based on print jobs. It differs from HCRP in that it needs no printer-specific drivers. This makes it more suitable for embedded devices such as mobile phones and digital cameras which cannot easily be updated with drivers dependent upon printer vendors.

Common ISDN Access Profile (CIP)

This provides unrestricted access to the services, data and signalling that ISDN offers.

Cordless Telephony Profile (CTP)

This is designed for cordless phones to work using Bluetooth. It is hoped that mobile phones could use a Bluetooth CTP gateway connected to a landline when within the home, and the mobile phone network when out of range. It is central to the Bluetooth SIG's '3-in-1 phone' use case.

Device ID Profile (DIP)

This profile allows a device to be identified above and beyond the limitations of the Device Class already available in Bluetooth. It enables identification of the manufacturer, product id, product version, and the version of the Device ID specification being met. It is useful in allowing a PC to identify a connecting device and download appropriate drivers. It enables similar applications to those the Plug-and-play specification allows.

Dial-up Networking Profile (DUN)

This profile provides a standard to access the Internet and other dial-up services over Bluetooth. The most common scenario is accessing the Internet from a laptop by dialing up on a mobile phone, wirelessly. It is based on Serial Port Profile (SPP), and provides for relatively easy conversion of existing products, through the many features that it has in common with the existing wired serial protocols for the same task. These include the AT command set specified in European Telecommunications Standards Institute (ETSI) 07.07, and Point-to-Point Protocol (PPP).

DUN distinguishes the initatior (DUN Terminal) of the connection and the provider (DUN Gateway) of the connection. The gateway provides a modem interface and establishes the connection to a PPP gateway. The terminal implements the usage of the modem and PPP protocol to establish the network connection. In standard phones, the gateway PPP functionality is usually implemented by the access point of the Telco provider. In "always on" smartphones, the PPP gateway is often provided by the phone and the terminal shares the connection.

Fax Profile (FAX)

This profile is intended to provide a well-defined interface between a mobile phone or fixed-line phone and a PC with Fax software installed. Support must be provided for ITU T.31 and / or ITU T.32 AT command sets as defined by ITU-T. Data and voice calls are not covered by this profile.

File Transfer Profile (FTP)

Provides the capability to browse, manipulate and transfer objects (files and folders) in an object store (file system) of another system. Uses GOEP as a basis.

Generic Audio/Video Distribution Profile (GAVDP)

Provides the basis for A2DP, and VDP.

Generic Access Profile (GAP)

Provides the basis for all other profiles. GAP defines how two Bluetooth units discover and establish a connection with each other.

Generic Attribute Profile (GATT)

Provides profile discovery and description services for Bluetooth Low Energy protocol. It defines how a set of ATT attributes are grouped together to form services.

Generic Object Exchange Profile (GOEP)

Provides a basis for other data profiles. Based on OBEX and sometimes referred to as such.

Hard Copy Cable Replacement Profile (HCRP)

This provides a simple wireless alternative to a cable connection between a device and a printer. Unfortunately it does not set a standard regarding the actual communications to the printer, so drivers are required specific to the printer model or range. This makes this profile less useful for embedded devices such as digital cameras and palmtops, as updating drivers can be problematic.

Health Device Profile (HDP)

Profile designed to facilitate transmission and reception of Medical Device data. The API's of this layer interact with the lower level Multi-Channel Adaptation Protocol (MCAP layer), but also perform SDP behavior to connect to remote HDP devices. Also makes use of the Device ID Profile (DIP).

Hands-Free Profile (HFP)

Currently in version 1.6, this is commonly used to allow car hands-free kits to communicate with mobile phones in the car. It commonly uses Synchronous Connection Oriented link (SCO) to carry a monaural audio channel with continuously variable slope delta modulation or pulse-code modulation, and with logarithmic a-law or μ-law quantization. Version 1.6 adds optional support for wide band speech with the mSBC codec, a 16 kHz monaural configuration of the SBC codec mandated by the A2DP profile.

In 2002 Audi, with the Audi A8, was the first motor vehicle manufacturer to install Bluetooth technology in a car, enabling the passenger to use a wireless in-car phone. The following year DaimlerChrysler and Acura introduced Bluetooth technology integration with the audio system as a standard feature in the third-generation Acura TL in a system dubbed HandsFree Link (HFL). Later, BMW added it as an option on its 1 Series, 3 Series, 5 Series, 7 Series and X5 vehicles. Since then, other manufacturers have followed suit, with many vehicles, including the Toyota Prius (since 2004), 2007 Toyota Camry, 2007 Infiniti G35, and the Lexus LS 430 (since 2004). Several Nissan models (Versa, X-Trail) include a built-in Bluetooth for the Technology option. Volvo started introducing support in some vehicles in 2007, and as of 2009 all Bluetooth-enabled vehicles support HFP.[2]

The Bluetooth car kits allow users with Bluetooth-equipped cell phones to make use of some of the phone's features, such as making calls, while the phone itself can be left in the user's pocket or hand bag. Companies like Johnson Controls, Peiker acustic, RAYTEL [3], Parrot SA, Novero, S1NN and Motorola manufacture Bluetooth hands-free car kits for well-known brand car manufacturers.

Most bluetooth headsets implement both Hands-Free Profile and Headset Profile, because of the extra features in HFP for use with a mobile phone, such as last number redial, call waiting and voice dialing.

Human Interface Device Profile (HID)

Provides support for devices such as mice, joysticks, keyboards, as well as sometimes providing support for simple buttons and indicators on other types of devices. It is designed to provide a low latency link, with low power requirements. PlayStation 3 controllers and Wii Remotes also use Bluetooth HID.

Bluetooth HID is a lightweight wrapper of the Human Interface Device protocol defined for USB. The use of the HID protocol simplifies host implementation (ex: support by Operating Systems) by enabling the re-use of some of the existing support for USB HID to also support Bluetooth HID.,

Headset Profile (HSP)

This is the most commonly used profile, providing support for the popular Bluetooth Headsets to be used with mobile phones. It relies on SCO for audio encoded in 64 kbit/s CVSD or PCM and a subset of AT commands from GSM 07.07 for minimal controls including the ability to ring, answer a call, hang up and adjust the volume.

Intercom Profile (ICP)

This is often referred to as the walkie-talkie profile. It is another TCS (Telephone Control protocol Specification)[4] based profile, relying on SCO to carry the audio. It is proposed to allow voice calls between two Bluetooth capable handsets, over Bluetooth.

LAN Access Profile (LAP)

LAN Access profile makes it possible for a Bluetooth device to access LAN, WAN or Internet via another device that has a physical connection to the network. It uses PPP over RFCOMM to establish connections. LAP also allows the device to join an ad-hoc Bluetooth network.

The LAN Access Profile has been replaced by the PAN profile in the Bluetooth specification.

Message Access Profile (MAP)

Message Access Profile (MAP)[5] specification allows exchange of messages between devices. Mostly used for automotive handsfree use. The MAP profile can also be used for other uses that require the exchange of messages between two devices. The automotive Hands-Free use case is where an on-board terminal device (typically an electronic device as a Car-Kit installed in the car) can talk via messaging capability to another communication device (typically a mobile phone). For example, Bluetooth MAP is used by HP Send and receive text (SMS) messages from a Palm/HP smartphone to an HP TouchPad tablet [6]. Bluetooth MAP is used by Ford in select SYNC Generation 1-equipped 2011 and 2012 vehicles [7] and also by BMW with many of their iDrive systems. The Lexus LX and GS 2013 models both also support MAP as does the Honda CRV 2012. Apple introduces Bluetooth MAP in IOS6.

OBject EXchange (OBEX)

See OBEX.

Object Push Profile (OPP)

A basic profile for sending "objects" such as pictures, virtual business cards, or appointment details. It is called push because the transfers are always instigated by the sender (client), not the receiver (server).

OPP uses the APIs of OBEX profile and the OBEX operations which are used in OPP are connect, disconnect, put, get and abort. By using these API the OPP layer will reside over OBEX and hence follow the specifications of the Bluetooth stack.

Personal Area Networking Profile (PAN)

This profile is intended to allow the use of Bluetooth Network Encapsulation Protocol on Layer 3 protocols for transport over a Bluetooth link.

Phone Book Access Profile (PBAP, PBA)

Phone Book Access (PBA)[8][9] or Phone Book Access Profile (PBAP) is a profile that allows exchange of Phone Book Objects between devices. It is likely to be used between a car kit and a mobile phone to:

- allow the car kit to display the name of the incoming caller;
- allow the car kit to download the phone book so the user can initiate a call from the car display.

The profile consists of two roles:

- PSE - Phone Book Server Equipment for the side delivering phonebook data, like a mobile phone
- PCE - Phone Book Client Equipment, for the device receiving this data, like a personal navigation device (PND)

Serial Port Profile (SPP)

This profile is based on ETSI 07.10 and the RFCOMM protocol. It emulates a serial cable to provide a simple substitute for existing RS-232, including the familiar control signals. It is the basis for DUN, FAX, HSP and AVRCP.

Service Discovery Application Profile (SDAP)

SDAP describes how an application should use SDP to discover services on a remote device. SDAP requires that any application be able to find out what services are available on any Bluetooth enabled device it connects to.

SIM Access Profile (SAP, SIM, rSAP)

This allows devices such as car phones with built-in GSM transceivers to connect to a SIM card in a phone with Bluetooth, thus the car phone itself doesn't require a separate SIM card. This profile is also known as rSAP (remote-SIM-Access-Profile). More information on which phones are supported can be found here(German version only)

Synchronization Profile (SYNCH)

This profile allows synchronization of Personal Information Manager (PIM) items. As this profile originated as part of the infrared specifications but has been adopted by the Bluetooth SIG to form part of the main Bluetooth specification, it is also commonly referred to as IrMC Synchronization.

Video Distribution Profile (VDP)

This profile allows the transport of a video stream. It could be used for streaming a recorded video from a PC media center to a portable player, or a live video from a digital video camera to a TV. Support for the H.263 baseline is mandatory. The MPEG-4 Visual Simple Profile, and H.263 profiles 3 and 8 are optionally supported, and covered in the specification.1

Wireless Application Protocol Bearer (WAPB)

This is a profile for carrying Wireless Application Protocol (WAP) over Point-to-Point Protocol over Bluetooth.

Comments

These profiles are still not finalised, but are currently proposed within the Bluetooth SIG:

- Unrestricted Digital Information (UDI)
- Extended Service discovery profile (ESDP)[10]
- Video Conferencing Profile (VCP) : This profile is to be compatible with 3G-324M, and support videoconferencing over a 3G high-speed connection.

Compatibility of products with profiles can be verified on the Bluetooth Qualification Program website [11].

References

[1] "Bluetooth Tutorial - Profiles" (http://www.palowireless.com/infotooth/tutorial/profiles.asp). palowireless Pty Ltd. . Retrieved 2007-01-05.
[2] "Bluetooth FAQ – Wireless In-Car Mobile Phone Technology – Volvo" (http://www.volvocars.com/au/sales-services/sales/car-devices/bluetooth/pages/faq.aspx). . Retrieved 27 Apr 2010. "phones must have the Bluetooth hands-free profile to be able to connect with the Volvo vehicle."
[3] http://www.egohandsfree.com
[4] "Bluetooth Glossary" (http://www.palowireless.com/infotooth/glossary.asp#SR). .
[5] Message Access Profile_SPEC_V10 (https://www.bluetooth.org/DocMan/handlers/DownloadDoc.ashx?doc_id=215400)
[6] Using Bluetooth wireless technology on a webOS device (http://kb.hpwebos.com/wps/portal/kb/na/touchpad/touchpad/wifi/solutions/article/21765_en.html)
[7] New SYNC Software Update Adds Bluetooth MAP Standard; Ford Poised to Give More Drivers Safer Texting Alternatives (http://corporate.ford.com/news-center/press-releases-detail/pr-new-sync-software-update-adds-35451)
[8] PBA - Phone Book Access (http://www.phonescoop.com/glossary/term.php?gid=323), Definition
[9] PBAP_SPEC_V11r00 (https://www.bluetooth.org/docman/handlers/downloaddoc.ashx?doc_id=230887)
[10] Extended Service Discovery Profile (http://www.palowireless.com/infotooth/tutorial/n6_esdp.asp)
[11] http://qualweb.bluetooth.org/Template2.cfm?LinkQualified=QualifiedProducts

External links

- Official website (http://www.bluetooth.com/)
- Specification: Adopted Documents (https://www.bluetooth.org/Technical/Specifications/adopted.htm), The Official Bluetooth SIG Member Website

Article Sources and Contributors

Nokia_Asha_302 *Source*: http://en.wikipedia.org/w/index.php?title=Nokia_Asha_302 *Contributors*: ChrisGualtieri, Cmg9751, Download, ItsMeOrYou, Jpoonnolly, Sanjeewa341, SudoGhost, W like wiki, Τασουλα, 18 anonymous edits

Series_40 *Source*: http://en.wikipedia.org/w/index.php?title=Series_40 *Contributors*: 1exec1, Agreimann, Alex Perry, Aniac, Ashrutisingh, Azuya, Bjelleklang, Blackvienna, Blakegripling ph, C933103, Calltech, Canaima, Centrx, Chapzboy, Charivari, Cnj, Daverocks, Davidbengtenglund, Diwas, Djlarz, Djr xi, Dktz, Dublinclontarf, Elspif, Ettrig, Evilboy, FleetCommand, Flod logic, Florin92, GRighta, Gaius Cornelius, Gaming&Computing, Hjorten, Imroy, ItsMeOrYou, JLaTondre, Jarekadam, Jasonon, Jesjimher, Joneilim, LordBen5000, MMuzammils, Mburdis, Mdwh, Milan292208, Monkeyhousetim, Msulik, Munamankeli, Park3r, Petri Krohn, Polisher of Cobwebs, Polluks, Rich Farmbrough, Rlobkovsky, Roleplayer, Rudiedude, Rwwww, SimonMenashy, Sofoklas, The Seventh Taylor, TheParanoidOne, Tirkfl, Txuspe, UKER, Vegaswikian, W like wiki, We hope, Xkury, Zakawer, ZeroOne, Zhernovoi, 106 anonymous edits

Nokia *Source*: http://en.wikipedia.org/w/index.php?title=Nokia *Contributors*: -Majestic-, 007zoo, 130.236.221.xxx, 159753, 16@r, 1exec1, 205ywmpq, 28421u2232nfenfcenc, A bit iffy, A box of sticks, A. B., A333, AB, ABF, AEMoreira042281, Abdul raja, Abnyc81, Academic Challenger, Accurate Nuanced Clear, Acdx, Acela Express, Adamrush, Adaobi, Adashiel, Adraeus, Aesopos, Aeusoes1, Agusm266, Ahazred8, Ahmedcena, Ahoerstemeier, Aijoovai, Aintneo, Alansohn, Aldis90, Alepik, Alex43223, Alexius08, AlfredWalsh, AliShaikh85, Alirezazzz, Alisha.4m, Allanvs, Allpower, Amitn, Andres, AndrewHowse, Andros 1337, Andyabides, Anir1uph, Anole 418, AnonMoos, Anshuln95, Antandrus, Antoncampos, Anubhavsharmaa, Apalsola, ApnAEA, Apparition11, Arch dude, Ariele, Armando, Artem-S-Tashkinov, Arthena, Arungm29, Arunsingh16, Asanka000, Ashmoo, Ashutosh.manager007, Atanasov, Avoided, Axeman89, BLAZEXBOY, BaSH PR0MPT, Backwalker, Badr55, Barek, Batbayarl, Beaker1306, Beao, Bearcat, Beetstra, Bender235, Benhocking, BernardZ, Bfaabaa, Big Brother 1984, BigT333, Billyboy1980, Blahma, Blake-, Blakegripling ph, Blobglob, Bloggert, Bluezy, Bmannaa, Bobblewik, Bobo192, Bogdangiusca, Bongomatic, Bongwarrior, Boomshadow, BorgHunter, BorisxXx, BoyanSyarov, Brennish, BrightStarSky, Brockert, BrokenSegue, Brokestudent007, Bruce1ee, BunnyT, C. A. Russell, C.Fred, C311u1ar, C628, CONFIQ, Cablecord, Cameron Scott, Can't sleep, clown will eat me, CanadianLinuxUser, Canterbury Tail, Capricorn42, Catiana 465, CecilWard, Cerebrith, Chamal N, Cheezy man, Cherkash, Cheung1303, Choptube, Chris the speller, ChrisHodgesUK, Chrisissocool, Chronulator, Clarkedexter, ClementSeveillac, CliffC, Clipmode, Closedmouth, Cmreditor, Cntras, Codey123, Coguar, Colibri37, Conversion script, Coolbull20, Coolslko, Cpl Syx, Crevox, Cryonic07, Crysb, Cybercobra, DBigXray, Dale Arnett, Damiens.rf, Damirgraffiti, Dancter, Daniel*D, Daniel575, DanielCD, Danim, Danio, DarkSaber2k, Darth Panda, Davidbspalding, Dazman 1988, Dck7777, Dearsina, Debresser, Delta avi delta, Den fjättrade ankan, Desaivishal14, Desbiadi, Diasimon2003, Dicklyon, Dicostathomas, Dima1, Discospinster, Dissolve, Diyar se, Djr xi, DmitTrix, DocendoDiscimus, Doctorcasey, DomQ, Dostal, Dostick, Download, Drewt, Drpickem, Dsavi.x4, Dyl, E Pluribus Anthony, E000xm, ES Vic, EWikist, Ed g2s, Editor182, Edsuom, Educatednawab, Edward, Eekerz, Eenu, Elfguy, Eliz81, Emmalewis1, Enemenemu, Epoxed, Eraserhead1, Erunestian, Esebi95, Estoy Aquí, Eta 94, Etincelles, Eurocanna, Everyking, FR Soliloquy, Falcon8765, Falconoffrance63, Faramir1138, Feezo, Felix Dance, Fffaaattt, Fieldday-sunday, Fixer88, FleetCommand, Flewis, Flora, Florentino floro, Florin92, Flrn, Flyguy649, Folksong, Fooishbar, Force39, Fraggle81, Fram, France64160, Franz-kafka, Fredrik, FreplySpang, Frymaster, FunkyDuffy, GMRE, Gachen, Gaius Cornelius, Galeon54, Galoubet, Gambit 28, Gandhietami, Gareth Griffith-Jones, Gaurav13dubey, Gavinio, Gdo01, Georgy90, Gethresh, Ghodannywahyudi99, Gimboid13, Gnangarra, Gnuton, Gobonobo, Godospoons, Gogo Dodo, Golemeye, Gr1st, Grafen, Graham87, Gratom, GreenJellybean, GreenJoe, Greenshed, Gregorydavid, Gringer, Groshna, Gsarwa, Guaka, Gump Stump, Gunglewack, Gunmetal Angel, Guy M, H.lloyd, H0dd0ck, Haakon, Hadal, Hankwang, Harishua, Harmi banik111, Harriv, HartzR, Hatesonyericsson, Hauskalainen, Havarhen, Hdt83, Head, Hell9, HelloAnnyong, Hemanshu, Heron, Herr Beethoven, HkCaGu, Hmains, Hooperbloob, Hotwiki, Howardchu, Htanna, Hu12, Hustedcarl, Hydrargyrum, Hylene, IJK Principle, Icewindfiresnow, Icseaturtles, Ilovemymac, Imgaril, Immunmotbluescreen, Imperi, Inspirito, Intelligentsium, Interframe, Inzy, Iohannes Animosus, Iphon, Irishnbears, Irstu, Isfisk, Ithinkicrappedmyself, Ixfd64, J. Sketter, J.delanoy, J36miles, JAAqqO, JCDenton2052, JForget, JGRIFF47, JIP, JLaTondre, JNW, Ja 62, Jackollie, Jak123, Jamcib, Jankratochvil, JayceAndTheNews, Jbsegal, Jclemens, Jdl18, Jeffrey Mall, Jeltz, Jensbn, Jeronimo, Jerry, Jerryseinfeld, Jim, Jimp, JmeSaunders, Jmh, Jni, JoeHinks, Joel7687, John, John Appleseed, JonForst6, Jonathan Hall, Jonik, Jopo, JorgeGG, Joseph Solis in Australia, Jovianeye, Jpk, Jrdioko, Jsysinc, JudasJesus, Julesd, Justin Steele, KC., KFP, KUsam, Kabsingh, Kai Ojima, Kalivd, Kallemax, Kalpesh.v.mistry, Kangaroopower, Katous1978, Kcandrsn, Keegan, Keonne, Ketchup, Khanri01, Kickus, Kirosana, Kjetilho, Kjramesh, Kkm010, Klilidiplomus, KoastalBenefitPromo, Konakalla anurag, Konrad Foerstner, Kosyoboy, Krellis, Kukulcan 7560, Kuponadam, Kwamikagami, LG4761, LLentil, Lafraia, Lamat, Larry laptop, LarryGilbert, Ld100, Leandrod, LeaveSleaves, Lectonar, Leo-loErlahell, Lepensky, Levineps, Lewys, Lfh, Lhotkami, Liamgilmartin, Lightmouse, LilHelpa, LittleDan, Loigenth, Lollomama, Look2See1, Loren.wilton, Lotje, Lovesameer9812, Lowflyingowl, Luigiacruz, Luisvmejia, Lukobe, M3lm4tt, MER-C, MMuzammils, Ma.abilash, Mabdul, Magioladitis, Makedonec28, Makeemlighter, Malhonen, Mani1, Manop, Marc Lacoste, Mardus, Marek69, Mark, Mark Arsten, Marmzok, Martarius, Martin.uecker, Mass09, Mastermind 147, Matrobriva, Matt povey, Mattbr, Mav, Maxim4o, Mayhaymate, Mayurg, McSly, McWika, Megahmad, Menphrad, Mert2000, MetroStar, Mhkay, Mic, Michaelbarr123, Microtony, Mike Rosoft, Mimsie, Minimac, Miraceti, Mirmo!, Misspsyb2, MithrandirAgain, Mkidson, Mmsteelers, Mojei, MominS, Momirt, MonaNL, Monkeynoze, Mowsbury, Mr Stephen, Mr. Met 13, MrFawwaz, MrOllie, MrZoolook, MrsAmethystWay, Mumble45, Murph146, Murphy418, Mysdaao, NKDurrani, Nagytibi, Nakon, Nantasatria, Narcisso, NawlinWiki, NeilN, Neilgravir, Neon white, Newone, Newtechpulse, Nielswik, Nikopolis1912, Ninja5624, Ninjustic, NokiaF, Nokiatrader, Noraft, Norden83, Nscheffey, Nubbly, Nubiatech, Number29, Nuno Tavares, Nuttycoconut, Nwpl, Obli, Ohnoitsjamie, OkakiMCMLXX, Oki putera, Oleksandr Kononenko, Oli Filth, Olivier, Omernos, One, Onlynokia, Onorem, Oo64eva, Ooza, Opraco, Ostralek, Otto2011, Oxymoron83, P0ppe, PTSE, Pablo-flores, Palefire, Parthrana, Pascal.Tesson, Pavel Vozenilek, PeeJay2K3, Pengo, Perohanych, Persia2, Pessi, PeteS, Peter Chastain, Petri Krohn, Pgan002, Pgk, Philip Trueman, Phonefinder2007, Pink Bull, Pista235, Pkadam, PkerUNO, Ploca12, Plokijnu, Pmggp, Pne, Pol098, Pony1142, Ppntori, Prakash.Akshat87, Prari, ProhibitOnions, Prokopov, Prolog, Proudfoot 001, Pseudomonas, Pterre, Puckly, Pudeo, Puneeeetjain, Pupster21, Quackdave, Quattrope, QuiteUnusual, Qwyrxian, Qxz, R'n'B, RTG, RVJ, Radagast83, RadicalBender, Ragityman, Rangoon11, Ratinator, Ratul655, Ravi vadgama, Ravo492, Rcawsey, RedWolf, Reliancepowercoin, Rent A Troop, Retired user 0001, Rettetast, Reuvengrish, RexNL, Rhobite, Richard Harvey, Ricktherazor, Riklear, Rjwilmsi, Rklawton, Roamataa, Rob Lindsey, Rob1974, Ron Ritzman, Ronnotel, Rrjanbiah, Rune X2, Ruronimomo, Ruslan0202, Rwalker, S h i v a (Visnu), S3000, SMC, SMP, ST47, Sadunlove, Sai2020, Saimhe, Salilm, Salome5764, Sam Hocevar, Samjuise, Samsara, Sander Säde, Sanixia, Sanjiv swarup, Satchmo2010, Savh, Savvo, Scootey, ScottSteiner, Scupplefish, Seanmilloy, Sebastian Shaw 449, Secaundis, Sedathut, Selimbey, Seqsea, Sfan00 IMG, Sfmammamia, Shadowjams, Sharcho, Shashankbhat, Shaun680, Shawnnicholsonca, Shd, Sherool, Shikker, Shizane, Siafu, SidP, Silpol, Silvah 87, SimonCrowley, SimonThird, Simone, Sinigagl, Skittle, Skyezx, Slavon37, Slo-mo, SmartFace44, SnappingTurtle, Snowolf, Sokratees9, Someguy1221, Soupyjnr, Spangineer, SpigotMap, Squash Racket, Squirtypants, Starblind, Steel, Stephan Leeds, Stephenb, SteveDay, SteveSims, Studioghiblitotoro, Suhailpeerbhai, Suomi Finland 2009, SuperHamster, Sven Manguard, Svgalbertian, Swctg, Syniq, T.O. Rainy Day, THEunique, TYelliot, Tagishsimon, TarzanASG, Tascha96, Tbhotch, Tda, Tedats, Teksosyete, Tellarin, Templetongore, Teqhed, Term061, Test2010, Thaliadrogna, The Man in Question, The Person Who Is Strange, The Random Editor, The Rogue Penguin, The Thing That Should Not Be, TheBlueKnight, TheGreenFaerae, TheYmode, Thebluebeast, Themfromspace, Thingg, Thorvy, Thule, Thunderbrand, TicketMan, Tide rolls, TigerK 69, Timo Honkasalo, Timonoko, Timster69, Tkynerd, Toehead2001, TomB123, Tomisti, Tompagenet, Tomwalden, Tonius, Tony1, Topperfalkon, Tpbradbury, Trakesht, Treschupetes, Tri400, Triage, Trisreed, Tsungik, Tuliopa, Turnstep, Twid, Ufinne, Ulric1313, Ultraviolet scissor flame, Ulysses, Uncle Dick, Unclealex, UpBeat, UpstateNYer, Valentine McKee, Varundbest10, Velella, Venus 9274, Versageek, Vespristiano, Vianello, Vina, Vinayakgole, Vinaywin7, Vininche, Vitund, Vkem, Volgar, Vrenator, Vuo, WIMYV, WJetChao, Wahgujarat, Waycool27, Weatherwax, Welovedoves, Weyes, Wfaulk, Wibbble, Wiki Wonda, Wikipelli, Wikizard2010, Wikizeta, William M. Connolley, WilliamKF, Wimt, WiseOne76, XLerate, Xhienne, Xyzahirm, Yakudza, Yamaaan, Yamamoto Ichiro, Yandman, Yanksox, YellowMonkey, Yousaf465, Yuhani, Yvresgyros, Z10x, ZZninepluralZalpha, Zackwee, Zeno Gantner, Zidonuke, ZirconiumTwice, Zodiak 887, Zpetro, Zsinj, Zunils, مجد ,امران ,المحارب ,محمد, 55דודו, [illegible], □ □ □ □ . □ □ □, 1422 anonymous edits

Wi-Fi *Source*: http://en.wikipedia.org/w/index.php?title=Wi-Fi *Contributors*: -Majestic-, 12dstring, 16@r, 19.158, 19.7, 220 of Borg, 2over0, 32deej, 802geek, A Softer Answer, A8UDI, A930913, ABF, AEMoreira042281, AGruntsJaggon, AL SAM, AWoodland, Aamrod, Aapo Laitinen, Abstract Idiot, Abune, Acebulf, ActivExpression, Adaclnews, Adashiel, Addihockey10, Addshore, Aeonsafe, Ahoerstemeier, Ajarmitage, Ajaysreedharan, Ajklein5211, Akerans, Akjar13, Alain r, Alan Liefting, Alansohn, Albedo, Albert109, Aldie, AlexF, AlexTiefling, Alexander UA, Alexius08, AlexiusHoratius, Algocu, AlistairMcMillan, AllanHainey, Allen Moore, Altaïr, Altermike, Alyssa hoffel, Amars, Amol.jawarkar10, Amorrow, AndersTR, Andrei Stroe, Andrew Maiman, Andy, Andy Dingley, AndyTheGrump, Andyiou52345, Andyjsmith, Andywmm9, Anna Lincoln, Anniex10, Anonymous anonymous, Anthony22, Antonrojo, Anttilk, Ardroliat, Aremith, Arichnad, Armando, Armyable, ArnoldReinhold, Arpabone, Arvindn, Asterion, Astral, Auric, Ausinha, Aveekbh, Awostrack, Awpsys, Baka toroi, Barek, BarretB, BartBenjamin, Bartman007, Baseball Bugs, Bawolff, Beavel, BelAir Networks Wireless Mesh, Beland, Belmond, Ben1220, Benhutchings, Beno1000, Berkut, Bevo, Bfigura's puppy, Bhny, Big Brother 1984, Bigbluefish, Bilbobee, BillC, Billywhack, Binkowski, Biscuittin, Blackdragon1157, Blackjewishfag, Blackroo1967, Blair Bonnett, Blooperfoob, Bluefoxicy, Bluezy, Bnordlund, Bobamnertiopsis, Bobblewik, Bobo192, Bonadea, Bongwarrior, Bornhj, Bozonz, Bratch, Brazil4Linux, Bremerenator, Briandjohnson, Brianga, BrightStarSky, Brillow, Broadband118, Brons, Brooknet, Btilm, BuddhaDharma, Bullzeye, Burnte, Bxmpls, C.Crane, C0nanPayne, C3k, CAJ, CGorman, CRH3 EMU, CUSENZA Mario, Cabalamat, Cailil, Calabe1992, Calabraxthis, Caliga10, Callidior, Calmer Waters, Caltas, Caltrop, Cambrant, Can't sleep, clown will eat me, CanisRufus, Capricorn42, CardinalDan, Carter, Casito, Catch2424, Cavanagh12345, Cbruno, Celtus, Charivari, Charles Moss, Charles Nguyen, Charm, Chaser, Chcknwnm, Chealer, Cheeesemonger, Chendy, Cherylcopeland, Chezi-Schlaff, Chitrapa, Chomperhead, Chowbok, Chris 73, Chris Q, Chrisjj, Christin varghese, Chunihan, Circeus, Classicrockfan42, ClementSeveillac, Click23, Clipdude, Closeapple, Closedmouth, Cmdrjameson, CnkALTDS, Coasterlover1994, CobraBK, Cocoaguy, Coffee, Coffin, Colenso, ColinHogben, Colinstu, Cometstyles, CommonsDelinker, Compellingelegance, Corevette, Corvus cornix, Courcelles, Cpl Syx, Crag, Crazycomputers, Crazyviolinist, Crimsondestroyer, Crissov, Cruisermjc, Cryout, Cshay, Cswilly, Cupcake1230, Cy21, CyberSkull, Cyrius, CzarB, D, DAJF, DARTH SIDIOUS 2, DMAch, DVdm, Dan Fuhry, DanPhilpott, Dancraggs, Daniel.Cardenas, DanielTahar, Dank, Danl999, Dannylim, Dark Shikari, Dark jedi requiem, Darkmaster2004, Darkwind, DarthRaider, DaveBurstein, Daven200520, Davhorn, DavidCary, Davidjk, Davidmack, Dawnseeker2000, DeVoteDz, DeadEyeArrow, Debresser, Deflective, Delirium, Dendodge, Denisarona, DerHexer, Devilalbert, Devin, Dieseldrinker, Dilane, Dilettante99, Dirkbb, Discospinster, DistopiaSPM, Djcapelis, Djg2006, Djkurtz, Dlohcierekim, Dmarquard, Dmccarty, Dmchyla, DocWatson42, Docboat, Dogman15, Domeilal, Doniago, DopefishJustin, Dopegirltwurk, Doyley, Dposse, Drakcap, Dreadstar, Dreamteamone, Drewzhrodague, Drplaster1, Ducknish, Dust Filter, Dylan Lake, DylanW, Dystopianray, Dzubint, E Wing, ESkog, EagleEye96, Earthlyreason, Eatsaq, Ecamdog, Eclectek, Ed Cormany, Ed g2s, Edcolins, Editor at Large, Eeekster, Eggnock, Egil, Ehn, Ejrs, Eken7, Elassint, Elbperle, Electro94, Electron9, Elvey, Em978, Enauspeaker, Engmark, Enviroboy, Epachamo, Epbr123, Eptin, Erencexor, Eric Kvaalen, Eric Wester, Eric-Wester, Ericyu, Eskalin, Espoo, Everyking, Evice, EvilPenguin, Excirial, Exit13, FF2010, FGrose, Falcon9x5, Faloopalump, FastLizard4, Favonian, Fbarton, Feedmecereal, Feureau, Fieldday-sunday, Fifieldt, Figueroaedgar, FilippoSidoti, Fireaxe888, Fiskbullar, Fitch, Fleisher, Fleminra, Flowerpotman, Fonzende, Fosnez, Foxj, Fractal3, Freakmighty, Frecklefoot, Freedomlinux, Freewol, Frencheigh, FreplySpang, Friginator, Ftbllfn33, Fubar Obfusco, Funandtrvl, Fuzheado, GGink, GNUtoo, GRAHAMUK, Gabbe, Gadfium, Gaius Cornelius, Galifardeu, Garion96, Gavinatkinson, Gazpacho, Gdo01, General Wesc, Genestr, Giftlite, Gilbertfein, Gilgamesh, Gilliam, Gimmetrow, Glacialfox, Glenn, GliderMaven, Gobbleswoggler, Godfather xie, Gojomo, Gokusandwich, Golbez, Golfandme, Gomm, Goodvac,

Gorank4, GozzoMan, Grafen, Grandfeller, Grandor, Greghe, Gregherlihy, Grim4593, Ground Zero, GroveGuy, Groxx, Guaka, Gulliveig, Gurch, Gutzmer, Guy Harris, Guycarmeli, Gwalarn, Gxti, Gyll, Gyrferret, H2g2bob, Hackfreewifi, Hadal, HaeB, Hagrinas, Hairy Dude, Halfbreath, Hankwang, Hanseichbaum, Haoie, Happenstancial, Happinessiseasy, Happysailor, HarisM, Harryzilber, Hawaiiboy99, Hede2000, Hedgehoglet, Helix84, Heman, HenkeB, Hera1aphrodite, Heron, Hertz1888, HexaChord, HiddenInPlainSight, Hillcrest, Holierthanthou, Holo16, Homerjay, Hsan22, Ht1848, Hu12, HybridBoy, Hydrargyrum, Hydroxides, I already forgot, INkubusse, IRP, IReceivedDeathThreats, Iainelder, IceKarma, Iflipti, Immunize, Impulse9, Inbamkumar86, Incompetence, Inkybutton, Inter16, Into The Fray, Intractable, Iridescent, Itai, Itusg15q4user, Ixfd64, Izzysanime, J.delanoy, JD554, JFreeman, JLaTondre, JOSamsung, JPMcGrath, Jafeluv, Jakohn, James Galloway, JamesAM, JamesBWatson, Jamesontai, Jamyskis, Jandalhandler, Janizary, Jarble, Jaryth000, Jason Recliner, Esq., Jason Stormchild, Jauricchio, Javawizard, Jaxsonjo, JayW, Jayharish, Jbowdenhm, Jbrandon2012, Jc4p, Jdlankin, Jeff G., Jeffrey Mall, Jeh, Jehochman, JeremyA, Jerome Charles Potts, Jesse Viviano, Jgamleus, Jh51681, Jhenderson777, Jim.henderson, Jingraham, Jjensen347, Jl1544, Jlmerrill, Jmlk17, Jnavas, Jncraton, JoGusto, JoanneB, John, John Mash, John254, JohnCD, JohnCub, Jon787, JonHarder, Jondel, JordoCo, Joseph Solis in Australia, Jossysayir, Joy, Joycloete, Jp78450, Jpatokal, Jtalledo, Juanscott, Jukeboxlord, Juliano, Justinko, Jvanthou, Jw21, Jwissick, KD5TVI, KGasso, Kantmorie, Katieh5584, Kautoorikrishna9, Kbrose, KeesCook, Keilana, KelleyCook, Kenyon, Kev19, Kevin chen2003, Kevinlie10, Kfluck, Kgopinath, Khalid Mahmood, Kikadeek, Kimiko, Kimvr, Kingboyk, Kingpin13, Kixy, Klingoncowboy4, KnowledgeOfSelf, Koman90, Kostisl, Kotra, Kozuch, Krellis, Kuru, Kvng, Kwamikagami, Kynn, Kzzl, Labboy, Laghlagh, Landroo, Larry laptop, LarryLACa, Laudaka, Lawnchair, Lawrence Cohen, Lcabanel, Le Fou, LeaveSleaves, Leendert, Leif, LeoDV, Leviwack, Lexor, Liam.winder@bozii.net, Liao, Lightmouse, Lights, Lilac Soul, Linesh1987, Linkspamremover, Lino Mastrodomenico, Lisatwo, Little Professor Stonecold, Little-man, Liyang, LizardJr8, Llort, Lmatt, Logan, Logical Cowboy, Logictheo, LordHector, LordOfer, Lucanos, Luci Sandor, Luis007cruz, Luk, M5, M7, MARQUIS111, MCB, MER-C, MLauba, MSTCrow, Mac, Maclaine, Madd the sane, Magioladitis, Magister Mathematicae, Mahendra, Mailer diablo, Mammad2002, Markaci, Markdr, Marketdiamond, Markpeak, Martinhayes, Master737373, Materialscientist, Matt Whyndham, Maxxdxx, Mazarin07, Mboverload, Mbw227, McSly, Mccombs1982, Mckenzie1995, Mcmarmatt, Me Three, Meco, MedicineMen, Meekywiki, Mehrunes Dagon, Melah Hashamaim, MennoMan, Merlion444, Metaclassing, Mgnbar, Mia.bo, Michael Hardy, Michael93555, MichaelMaggs, Michaelsbll, Microfrost, Mifter, Mike Moreton, Mike.lifeguard, MikeFenney, MikeM8892, Mikemoral, Mindmatrix, Miranda, Miremare, Mirko.fidisk.fi, Misternuvistor, Mjmarcus, Mkzz, Mmdoogie, Mmmcatsoup, Moarmudkipzlulz, MobileMistress, Modulatum, Moehrchen, Mollykate82, Momo san, Monaarora84, Monkeyman, Monochnotos, Monuko, Moreati, Mormegil, Mortense, Mothmolevna, Mowgli, Moyda, Mrmiscellanious, Mshahbaig, Msm20032003, Mtonkin, Mumia-w-18, Mushroom, Muzzle, Mwanner, Mygerardromance, Mytimekept, Namenad, Nankeyman, Nasa-verve, Nbarth, Ncmvocalist, Ndyguy, Nealcardwell, NellieBly, Nelson50, NeoChaosX, Nerminbarman, Networkingguy, Neurolanis, Niemeyerstein en, Night Gyr, Nil0lab, Niri.M, Nixdorf, Njh@bandsman.co.uk, Noah Salzman, Noctibus, Nopetro, Notedgrant, Novalis, Nowcrash, Nposs, Nsaa, Nux, OPless, Obastani, Octahedron80, Ohnoitsjamie, Oldr4ver, Omegatron, Omhafeieio, Omicronpersei8, Onorem, Otivaeey, Oxymoron83, Oystein, Ozseden, Pacific1982, Paddu, PamD, Parrotheadmjb, Parsmutaf, ParticleMan, Patrick, Patrick Fisse, Patstuart, Paul Weaver, Paulshanks, Peak, Pekaje, Pelago, PeregrineAY, Peter, Peter McGinley, Peterl, Petrb, Petri Krohn, Peyre, Phatom87, Phi beta, Phil-welch, PhilKnight, Philip Trueman, Phlegat, Phooto, Photocopier, Phreakuency, Piano non troppo, Pie4all88, Pieleric, Piet Delport, Pimfig, PinkPig, Pip2andahalf, PleaseInsertGirder, PoTi, Pokipsy76, Pol098, Polaralex, Porschedriver403, Preslethe, PrestonH, Primary0, ProhibitOnions, Project2501a, Pt, Public Menace, Puggs, Pujanmalla, Puneet.kasera, Quantpole, Qui1che, Quintote, QuiteUnusual, Qutezuce, R6144, RA0808, RDOlivaw, RHaworth, Radagast, Radiant chains, RadioActive, Raj bhinde, Rannpháirtí anaithnid, Ravendrop, Ravenperch, Raxian, Ray Radlein, Raysonho, Raywil, Rcassidy, Rce-revo, Reach Out to the Truth, Rearden9, Recognizance, Reconsider the static, Recurring9, Rednectar.chris, Rees11, Regara MkII, Reisio, Rememberway, Renewal36, RenniePet, Requestion, Res2216firestar, Retired username, Retodon8, Rewt, Rfc1394, Rhobite, Riadlem, Ribread2, Rich Farmbrough, Rich257, Richwales, Ricksterrr11, Riddz17, Ripetom, Risk one, Riskariandita, Rjanag, Rjhatl, Rjwilmsi, Rkopplin, Rmallins, Robartin, Robert Cassidy, Robertvan1, Robo56, RockOfVictory, Rogerbrent, Rojypala, Romanski, Rootxploit, Roscelese, RoseTech, Roybadami, Rror, Rsrikanth05, Rtrohantyagi, RufusThorne, Rumping, RupertMillard, Russoc4, Ryan Postlethwaite, RyanCross, Rynsaha, Ryuch, S kollmor, S.K., S.ferguson, SEWilco, SF007, SHCarter, SMC, SPANGER BANGER, STGM, Saikiri, Sailor iain, Salty!, Sam Hocevar, Sam Korn, Sandeepreddyus, Sander123, Sandstein, Sango123, Sastrycpps, Satishchitti02, Saxtonrob, Sbenton, Sbmeirow, Sburpeeduncan, Sceptre, Scholia, SciCorrector, Sckirklan, Scurless, Seadood90, Seajay, Sean Reynolds, Sean2074, Seancoady, Seans Potato Business, Seaphoto, Searchme, Secfan, Seikku Kaita, Seqsea, Sesu Prime, Setveen, Sevenneed, Shanafme, Shandris, Shinpah1, Shirik, Shorty114, Shuipzv3, Shunt010, Sikon, Silverxxx, Silvery, Simm, SimonP, Simonn, SineWave, Sixteen Left, Sjjupadhyay, Skatebiker, Skategem, Skierpage, Skor, Slippyd, Slon02, Slowking Man, Smailis, Smalljim, Smoke, Snori, Snowolf, SoCalSuperEagle, Soap, Socrates2008, Sodapop89, Softlavender, Sonam.bhutia, SonicBlue, Soumyasch, Souseiseki42, SpacedOut, Spathak112, Spearhead, Speck-Made, Spitfire19, Spookee, SpuriousQ, Sraraja, Srborlongan, Srikeit, Ssd, Ssri1983, Stangaa, Star Trek Man, StaticGull, Stephan Leeds, Stephen, Stephenb, Stevenj, Stickee, Styrofoam1994, Suid-Afrikaanse, Sunray, SuperDude115, Superborsuk, Supermanden, Superway25, SusanLesch, Susvolans, Suwa, Svetovid, Swizman, T callahan, T h e M a v e r i c k, Tabledhote, Tainter, TakuyaMurata, Taralsoni, TastyPoutine, Tbhotch, Tbkexile, Tcncv, Teapeat, Teddybearspicnic, Teixant, Tekinera, Teles, Texaswebscout, Texture, Tfl, Thatguy553, The Belgain, The Bryce, The Mark of the Beast, The Rambling Man, The Thing That Should Not Be, The Utterly Annoying Pedant, The undertow, The wub, TheGWO, TheLiberalTruth, TheRealFennShysa, TheTito, TheWishy, Thentukiran, Thepidding, Thetasig, Theymos, Thiago R Ramos, Thingg, Think outside the box, Think smith, ThrashedParanoid, Thunderbrand, Tide rolls, TigerShark, Tim1988, Timbatron, Timeshift9, TippyGoomba, Tiptoety, Tjrana0, Tntnnbltn, Tobiaslw, Toccatafugue, Tohd8BohaithuGh1, Tokachu, Tom.k, Tommy2010, Tony Sidaway, Tony1, Topazg, Tpbradbury, Tree Biting Conspiracy, Tri400, Trlkly, Trusilver, Tsange, Tsenapathy, Ttiotsw, Tucker001, TucsonDavid, Twang, Twasono, Twocs, Twp, Tyler, UU, Ultraman2008, Umerqureshi, Unprovoked, Unused0030, Useight, Utopianfiat, Valfontis, Valley2city, Vedantm, Vegaswikian, Veledan, Verbal, Versageek, Versus22, Vgautham 91, Viajero, Vibhijain, Vicarious, Vickymesingh, Viktor Laszlo, Vinaywin7, Vishteen, VodkaJazz, Vossman, Vrenator, W Nowicki, WLRoss, Waggers, Wahrmund, Walloon, Walter Görlitz, Wandab12, Warwickcaddie, WatermelonPotion, Watkinsclass67, Wavelength, Wdfarmer, Webwizard, Wehe, Wellithy, Wellivea1, Wendell, Wep, Wereon, West.andrew.g, Whitepaw, Wifieverywhere2007, Wikcerize, Wikipedian314, Wikipelli, Wikiuser100, Wikizeta2, William Avery, WilliamH, Wimt, Wingspantt, Wizardist, Wjejskenewr, Wknight94, Wo.luren, Wog7777, Wolfkeeper, Woohookitty, Wormeyman, Wrrr, Wskish, Wtshymanski, Ww, Wysprgr2005, XSSX, Xmnemonic, Xtefan, Xtifr, XxxFilipks, Yosri, Your wasted elf, Yuhong, ZKPilot, Zach Vega, Zack Holly Venturi, Zarherif, Zeroshell, Zgreycoat, Zzuuzz, স্বপ্ন, ന്യൂ, පසිදු කාවින්ද, □ □ , □ □ □ □ □ □ , 2874 anonymous edits

Symbian *Source*: http://en.wikipedia.org/w/index.php?title=Symbian *Contributors*: 1exec1, AJ-India, Aednichols, AgarwalSumeet, Ahmadadam96, Airplaneman, Ajay vishvanathan, Andries, Anujgupta2 979, Apdevries, Arbitrarily0, Armando, Arnaud92, Asiaworldcity, Atlan, Auntof6, Aymen ka, B0o-supermario, BD2412, BSCorrector, Bajjibala, Belastro, Binoyjsdk, Blakegripling ph, Bpringlemeir, Brianreading, BrideOfKripkenstein, Bruce1ee, C933103, Cameron Scott, Capdor, CapitalLetterBeginning, Cayun, Ceancata, Cherkash, Chris the speller, Clsin, CommonsDelinker, Conpibien, Danhash, Daniel.Cardenas, DarwinPeacock, Davidjcole, Dian Putrawangsa, Digitalsurgeon, DmitryKo, DmitryRodriguez, Editor182, Edward, Elephant in a tornado, Empty Buffer, Erianna, ErikBatu, Favonian, Fences and windows, Ferrero777, Fetchcomms, Ffaarrooqq, FinnsDeal, FireyFly, FleetCommand, Florin92, Fraggle81, Freddiegjertsen, Fru1tbat, Gaming&Computing, Garyp01, Gene91, Gerhman, Gmihaylo, Gryllida, Gsarwa, Haakon, Hersfold, Hoshie, Hu12, Hydrox, IJK Principle, Imcdnzl, Intractable, IsmaelLuceno, Iuhkjhk87y678, JHP, JaGa, Jacek Kendysz, Jaizovic, Jayantanth, Jeffrey04, Jerryobject, Jfourgeaud, Jfromcanada, Jhertel, Jondoelocksmith, Jwoodger, Khanayub1986, Kinema, Kiran Gopi, Kkm010, Knkw, Knurdtech, Koavf, Krashlandon, KrisBogdanov, Krischik, KumardipSarkar, Kyrios320, LemChops, Lester, Ligertail, LokiClock, Lori-m, Ltomuta, Magister Mathematicae, Magog the Ogre, Mardus, MargaritMargaryan, Martarius, Matemaciek, Materialscientist, Mattkap2, Mdwh, Meewam, Mephistophelian, Micahblitz, MichaelBillington, Mikelyons, Mild Bill Hiccup, Minikola, Mnia786, Mrprajesh, Muhandes, Mushroom, Mushroom9, Myscrnnm, Narasimhanator, Nate1481, NotWith, Notedgrant, Obiwankenobi, Oknazevad, OlEnglish, OwenBlacker, Parthibls, Pnm, Pol098, Polluks, Ponydepression, Quietbritishjim, R'n'B, Rabit gti, Reboot, Rjd0060, Sainath468, Seth Nimbosa, Sharcho, Shirudo, Shrish, Smileverse, SoWhy, Soulxlight, Soumilj, Stichbury, Surya Prakash.S.A., Svick, TRS80Math, Tahir mq, Tgeairn, Thumperward, Timotheus Canens, Tirim4, Tomjenkins52, Traveler100, Trusilver, Truth Lover80, UKER, Ufim, VMlemon, Vervamon, VijayPadiyar, Vishnunj, Wattyirl, Weeder 001, Why vincent, Yavoh, Yorxs, Zackaback, Zagothal, ZeroOne, Zhongle, ZirconiumTwice, Zvrkljati, ಮಹೇ, 303 anonymous edits

Broadcom *Source*: http://en.wikipedia.org/w/index.php?title=Broadcom *Contributors*: Adraeus, Alecv, Alfski, Amcl, Angela, AnthonyWong, Astor14, Atharkhan, Auntof6, Blaxthos, Bobbyfisherman123, Bongomatic, Boothy443, BsdIsDying, Btilm, Buchanan-Hermit, CDA, Carlossuarez46, Catapult, Chaitanya.lala, ChrisGualtieri, ChrisRuvolo, Cocoaguy, Colonies Chris, Coolcaesar, Cootiequits, Cupfullofpens, Curtmcd, D235j, Dananimal, Danausi, Dannycrouch, Dbiagioli, Deineka, DevAnubis, Dewey Finn, Diablo-fan, Do it, Dreemteem, Dyl, EWS23, Egil, EricEnfermero, Eugrus, Flibble, FluffyWhiteCat, Frap, Gala.martin, Galoiserdos, Geek2003, Gogo Dodo, Goldsteinal, Greenshed, Ground Zero, Gsarwa, Guy Harris, Gwernol, HangingCurve, IRWolfie-, Idaltu, Inzider, Jalanpalmer, Jamcib, Jehochman, JohnSRussell3Finley, Jondel, Jovianeye, Jvcdude, Jwojdylo, Karebear 1022, KeepingPace, Keith Edkins, Kenyon, Khfan93, Klemen Kocjancic, Kozuch, Lightmouse, Lipatden, Lotje, Mac, Malepheasant, Margin1522, Matt.Messina, Mehudson1, Mektroid, Mohawk11111, MrRadioGuy, MurryUp, Mushroom, Neuro, Ocedits, PacificJuls, Paragdighe, PhilipO, Phoneyplayer, Psantora, Quadra630, Ralphbk, Rangoon11, Ray99wik, Renegadeviking, Rich Farmbrough, Rjwilmsi, RobyWayne, RyanQuinlan, RzR, SF007, Samcgregor, Sdiazbrcm, Seaphoto, Semiexpert, Skizzik, Smmgeek, Solarisworld, Sterling63, TechGeek70, Tedder, The wub, Thepangelinanpost, Tinton5, Tonkie, Tregoweth, Triikan, Tsunanet, Turkeyphant, Vanderesch, Vargklo, Vedant lath, Vegaswikian, Vinayextc, Voxhumana, W Nowicki, WhisperToMe, Whitethunder79, Xnatedawgx, YouTubeFan123, Zian, 183 anonymous edits

USB_On-The-Go *Source*: http://en.wikipedia.org/w/index.php?title=USB_On-The-Go *Contributors*: Ahruman, AlexChurchill, AlistairMcMillan, Andries, Andy W Mitchell, Asiir, Bgkwtnyqhzor, Billgordon1099, Damian Yerrick, Danhash, DanielGr, DarkSTALKER, David Gerard, David Haslam, Debresser, Diamondland, Differo, Electron9, Frap, Geniac, Gistereziz, Gronky, Gurch, Harvester, Hecline3rd, Heron, Hunding, Isnow, Itsacon, JCLAWSON, Jeff Wheeler, Jeffthejiff, Jerryobject, Justinz30, Kakoui, Kozuch, LeadSongDog, Lexein, Macskull, Malinthasa, MarkRodda, Milominderbinder2, Mortense, Mrzaius, Nicolas.brouard@gmail.com, Nuno Tavares, OS2Warp, Oleh Kernytskyi, Pelago, Plasticup, Pmc, Privateboz, RainbowOfLight, Rchandra, Requestion, Retodon8, Rjpetrie, Roger Fiander, SebastianBreier, SivaKumar, Snori, Sparklex, Spidey104, Stephan Leeds, TheMagician, UKER, Velella, Vjdchauhan, Voidxor, Why Not A Duck, Wsmarz, X96lee15, Ytgy111, Zaphodikus, 75 anonymous edits

Bluetooth_profile *Source*: http://en.wikipedia.org/w/index.php?title=Bluetooth_profile *Contributors*: Abdull, AndrewHowse, Androux, Aninhumer, Atomes, Avernet, Babca, Bananabender, Bollyjeff, Capricorn42, Cburnett, Cfp, CheShA, Chris Murphy, Chris the speller, Closeapple, Coastergeekperson04, Cromeyer, D2brothe, DC, Daniel 1992, DavesPlanet, Destynova, Dingledow, Discpad, Dj stone, Doniago, Dwarak24, EdoDodo, Edoe, Elikser, EnTerr, Erechtheus, Eskdale, FLaSh PT, Fabianseitenblick, Gioto, Glenn, GrayFullbuster, Ground Zero, Hoene, Horcrux92, Hqb, Hut 8.5, Inkling, Inspector 34, Intgr, J.delanoy, JHunterJ, Jarl Friis, Jeroenco, Jk2q3jrklse, Jlarkin, Jmrowland, Jni, Jpp, Jsharpminor, Jupter-manzana, KDesk, Keumkang, Kusunose, Lantrix, Lathe26, Lbecque, Lightmouse, Luk, M5, Mac, Mandarax, Margol, Mark Shelby, Martin S Taylor, Miahfost, MichaelStanford, MikeCerm, Mikeo, Mild Bill Hiccup, Miquonranger03, Mount Alverno, Nihiltres, Nihonjoe, Nopetro, Nordicflow, OS2Warp, Ohnoitsjamie, Ouro, Pb30, Pbryan, Perennialmind, Peterpall, Petree, Phatom87, Philipkoo2, Pinethicket, Pok148, Potatoswatter, Psantora, RT100, RatherJovialTim, Rbrewer42, Revzoom, Ricgal, Rjwilmsi, Rob Blanco, RudolfSimon, Rwestafer, SMcCandlish, ScottJ, Silas S. Brown, Squeakypaul, Stechert, Stephan Leeds, Stevan White, Stjson, TheBrave, Theomamentalist, TianzhouChen, TigerK 69, TyA, VDD139, Velella, Versageek, Vindictive Warrior, Wafry, Webhat, Whitepaw, Wikibarista, Wikihd, William M. Connolley, WinTakeAll, Wireless friend, Woohookitty, WurmWoode, Xpclient, Yannick56, Zabacad, 274 anonymous edits

Image Sources, Licenses and Contributors

Image:Nokia 6300.jpg *Source*: http://en.wikipedia.org/w/index.php?title=File:Nokia_6300.jpg *License*: Public Domain *Contributors*: Original uploader was Racklever at en.wikipedia

File:Symbian S40 v10.80.png *Source*: http://en.wikipedia.org/w/index.php?title=File:Symbian_S40_v10.80.png *License*: Creative Commons Attribution-Sharealike 3.0 *Contributors*: Blackvienna, 1 anonymous edits

File:Nokia wordmark.svg *Source*: http://en.wikipedia.org/w/index.php?title=File:Nokia_wordmark.svg *License*: Trademarked *Contributors*: Apalsola, Bencmq, ELeschev, Editor182, Hautala, Yarl, 1 anonymous edits

File:Decrease2.svg *Source*: http://en.wikipedia.org/w/index.php?title=File:Decrease2.svg *License*: Public Domain *Contributors*: Sarang

Image:Fredrik Idestam.png *Source*: http://en.wikipedia.org/w/index.php?title=File:Fredrik_Idestam.png *License*: Public Domain *Contributors*: style="background: #E4E4E4; color: black; vertical-align: middle; text-align: center; " class="unknown table-unknown"|Author

Image:Leo Mechelin (cropped).png *Source*: http://en.wikipedia.org/w/index.php?title=File:Leo_Mechelin_(cropped).png *License*: Public Domain *Contributors*: style="background: #E4E4E4; color: black; vertical-align: middle; text-align: center; " class="unknown table-unknown"|Author

File:Nokia 150 and nokia 1100.jpg *Source*: http://en.wikipedia.org/w/index.php?title=File:Nokia_150_and_nokia_1100.jpg *License*: Public Domain *Contributors*: Lvova Anastasiya (Львова Анастасия, Lvova)

File:Nokia booklet 3g-10 (3949263497).jpg *Source*: http://en.wikipedia.org/w/index.php?title=File:Nokia_booklet_3g-10_(3949263497).jpg *License*: Creative Commons Attribution-Sharealike 3.0 *Contributors*: http://thenokiablog.com/Reposted by Mark Guim from United States

File:Nokia HQ.jpg *Source*: http://en.wikipedia.org/w/index.php?title=File:Nokia_HQ.jpg *License*: Creative Commons Attribution-Sharealike 3.0,2.5,2.0,1.0 *Contributors*: -Majestic-

File:Nokia evolucion tamaño.jpg *Source*: http://en.wikipedia.org/w/index.php?title=File:Nokia_evolucion_tamaño.jpg *License*: Public Domain *Contributors*: Jorge Barrios

File:All 9xxx.png *Source*: http://en.wikipedia.org/w/index.php?title=File:All_9xxx.png *License*: GNU Free Documentation License *Contributors*: derivative work: -Majestic- (talk) All9xxx.jpg: Original uploader was R@y at de.wikipedia (2004-04-03)

File:Nokia N8 (front view).jpg *Source*: http://en.wikipedia.org/w/index.php?title=File:Nokia_N8_(front_view).jpg *License*: Public Domain *Contributors*: Editor182, Vivinnl, X-Pilot

File:SymbianWMWP7USMarketShare.png *Source*: http://en.wikipedia.org/w/index.php?title=File:SymbianWMWP7USMarketShare.png *License*: Creative Commons Attribution-Sharealike 3.0 *Contributors*: User:Enemenemu

File:Nokia E55 01.jpg *Source*: http://en.wikipedia.org/w/index.php?title=File:Nokia_E55_01.jpg *License*: Creative Commons Attribution-Sharealike 2.0 *Contributors*: James Nash

File:Nokia N900 1.jpg *Source*: http://en.wikipedia.org/w/index.php?title=File:Nokia_N900-1.jpg *License*: Creative Commons Attribution-Sharealike 3.0,2.5,2.0,1.0 *Contributors*: User:Ilya Voyager

File:Nokia E90 communicator.JPG *Source*: http://en.wikipedia.org/w/index.php?title=File:Nokia_E90_communicator.JPG *License*: Creative Commons Attribution 3.0 *Contributors*: Georgy90

File:Nokia5800xpress.png *Source*: http://en.wikipedia.org/w/index.php?title=File:Nokia5800xpress.png *License*: Creative Commons Attribution-Sharealike 3.0 *Contributors*: Nokia_5800_XpressMusic_Browser.jpg: Jupter-manzana derivative work: TheAdam0s (talk)

File:Flag of Canada.svg *Source*: http://en.wikipedia.org/w/index.php?title=File:Flag_of_Canada.svg *License*: Public Domain *Contributors*: Anomie

File:Flag of Finland.svg *Source*: http://en.wikipedia.org/w/index.php?title=File:Flag_of_Finland.svg *License*: Public Domain *Contributors*: Drawn by User:SKopp

File:Flag of the United States.svg *Source*: http://en.wikipedia.org/w/index.php?title=File:Flag_of_the_United_States.svg *License*: Public Domain *Contributors*: Anomie

File:Flag of Switzerland.svg *Source*: http://en.wikipedia.org/w/index.php?title=File:Flag_of_Switzerland.svg *License*: Public Domain *Contributors*: User:Marc Mongenet Credits: User:-xfi- User:Zscout370

File:Flag of Germany.svg *Source*: http://en.wikipedia.org/w/index.php?title=File:Flag_of_Germany.svg *License*: Public Domain *Contributors*: Anomie

File:Flag of Norway.svg *Source*: http://en.wikipedia.org/w/index.php?title=File:Flag_of_Norway.svg *License*: Public Domain *Contributors*: Dbenbenn

File:Flag of France.svg *Source*: http://en.wikipedia.org/w/index.php?title=File:Flag_of_France.svg *License*: Public Domain *Contributors*: Anomie

File:Nokia Connecting People.svg *Source*: http://en.wikipedia.org/w/index.php?title=File:Nokia_Connecting_People.svg *License*: Public Domain *Contributors*: Original uploader was -Majestic- at en.wikipedia

File:Nokian logo.svg *Source*: http://en.wikipedia.org/w/index.php?title=File:Nokian_logo.svg *License*: Public Domain *Contributors*: Z10x at en.wikipedia

File:Nokian pääkonttori Keilaniemessä.jpg *Source*: http://en.wikipedia.org/w/index.php?title=File:Nokian_pääkonttori_Keilaniemessä.jpg *License*: GNU Free Documentation License *Contributors*: J-P Kärnä

File:Wi-Fi Logo.svg *Source*: http://en.wikipedia.org/w/index.php?title=File:Wi-Fi_Logo.svg *License*: Public Domain *Contributors*: Wi-Fi Alliance

File:Loudspeaker.svg *Source*: http://en.wikipedia.org/w/index.php?title=File:Loudspeaker.svg *License*: Public Domain *Contributors*: Bayo, Gmaxwell, Husky, Iamunknown, Mirithing, Myself488, Nethac DIU, Omegatron, Rocket000, The Evil IP address, Wouterhagens, 20 anonymous edits

File:Wifi logo.jpg *Source*: http://en.wikipedia.org/w/index.php?title=File:Wifi_logo.jpg *License*: Creative Commons Attribution-Sharealike 3.0 *Contributors*: Krish Dulal

File:Metro Wireless Node.jpg *Source*: http://en.wikipedia.org/w/index.php?title=File:Metro_Wireless_Node.jpg *License*: Creative Commons Attribution-Sharealike 3.0 *Contributors*: Robo56

File:WiFi-detector.jpg *Source*: http://en.wikipedia.org/w/index.php?title=File:WiFi-detector.jpg *License*: Public Domain *Contributors*: Raysonho@Open Grid Scheduler

File:RouterBoard 112 with U.FL-RSMA pigtail and R52 miniPCI Wi-Fi card.jpg *Source*: http://en.wikipedia.org/w/index.php?title=File:RouterBoard_112_with_U.FL-RSMA_pigtail_and_R52_miniPCI_Wi-Fi_card.jpg *License*: GNU Free Documentation License *Contributors*: Kozuch

File:3GN .jpg *Source*: http://en.wikipedia.org/w/index.php?title=File:3GN_.jpg *License*: Public Domain *Contributors*: Original uploader was Rewt at en.wikipedia

File:EBWifi.JPG *Source*: http://en.wikipedia.org/w/index.php?title=File:EBWifi.JPG *License*: Creative Commons Attribution-Sharealike 3.0 *Contributors*: highwycombe (talk)

File:Wireless adaptor USB.jpg *Source*: http://en.wikipedia.org/w/index.php?title=File:Wireless_adaptor_USB.jpg *License*: Creative Commons Attribution 3.0 *Contributors*: Ben Ben, Silverxxx, 1 anonymous edits

File:Ezurio wism2 small.jpg *Source*: http://en.wikipedia.org/w/index.php?title=File:Ezurio_wism2_small.jpg *License*: GNU Free Documentation License *Contributors*: Phooto, 4 anonymous edits

File:Nokia C7 with Nokia Belle.jpg *Source*: http://en.wikipedia.org/w/index.php?title=File:Nokia_C7_with_Nokia_Belle.jpg *License*: Creative Commons Attribution-Sharealike 3.0 *Contributors*: Seqtre

File:Increase2.svg *Source*: http://en.wikipedia.org/w/index.php?title=File:Increase2.svg *License*: Public Domain *Contributors*: Sarang

Image:Broadcomheadquarters.jpg *Source*: http://en.wikipedia.org/w/index.php?title=File:Broadcomheadquarters.jpg *License*: GNU Free Documentation License *Contributors*: en:User:Coolcaesar

CPSIA information can be obtained at www.ICGtesting.com
Printed in the USA
LVOW061141210113
316539LV00005B/703/P